CONTEMPORARY TOPICS
IN IMMUNOCHEMISTRY
VOLUME 1

CONTEMPORARY TOPICS IN IMMUNOCHEMISTRY

CONTEMPORARY TOPICS IN IMMUNOCHEMISTRY

VOLUME 1

EDITED BY F. P. INMAN

Department of Microbiology
The University of Georgia
Athens, Georgia

SPRINGER SCIENCE+BUSINESS MEDIA, LLC

Library of Congress Catalog Card Number 73-186260
ISBN 978-1-4757-1345-9   ISBN 978-1-4757-1343-5 (eBook)
DOI 10.1007/978-1-4757-1343-5

© 1972 Springer Science+Business Media New York
Originally published by Plenum Press, New York in 1972
Softcover reprint of the hardcover 1st edition 1972

CONTRIBUTORS TO THIS VOLUME

Gordon L. Ada   *Department of Microbiology, John Curtin School of Medical Research, Australian National University, Canberra, Australia*

E. Benjamini   *Department of Medical Microbiology, School of Medicine, University of California, Davis, California*

Joel W. Goodman   *Department of Microbiology, University of California, San Francisco, California*

Barry D. Kahan   *Department of Surgery, Massachusetts General Hospital, Boston, Massachusetts*

Richard A. Lerner   *Department of Experimental Pathology, Scripps Clinic and Research Foundation, La Jolla, California*

Christopher R. Parish   *Department of Microbiology, John Curtin School of Medical Research, Australian National University, Canberra, Australia*

R. R. Porter   *Department of Biochemistry, University of Oxford, Oxford, England*

Ralph A. Reisfeld   *Department of Experimental Pathology, Scripps Clinic and Research Foundation, La Jolla, California*

R. J. Scibienski   *Department of Medical Microbiology, School of Medicine, University of California, Davis, California*

Hans L. Spiegelberg   *Department of Experimental Pathology, Scripps Clinic and Research Foundation, La Jolla, California*

K. Thompson   *Department of Medical Microbiology, School of Medicine, University of California, Davis, California*

Preface

In 1897, Ehrlich suggested that natural preformed receptors from the surface of cells provided immunity to various chemical substances. Many years later, in 1940, Pauling proffered the concept that antibodies comprised a single polypeptide chain and that each end of the protein could form an antigen-binding site. Burnet tried to explain the diversity of antibody specificity by hypothesizing that it was cell-derived. These hypotheses probably have led to as much or more experimentation and discussion than any other of the many conjectures set forth to explain immunity on a molecular and cellular basis.

Extensive investigations, initially stimulated by these propositions, proved Pauling's notion incorrect. In its demise, however, the multichain structure of the immunoglobulins was realized. In retrospect it becomes obvious that Ehrlich's idea, though not correct, was borne of amazing logic and cognition. Expansion of Burnet's theory seems to be occurring presently; much excitement is engendered by the finding of cell-bound immunoglobulin receptors.

During the preceding dozen years, immunochemists have accumulated enormous quantities of data. Though there is so much yet to be done, as a result of this research one may now discuss antigen-binding sites in relation to the protein's primary structure. There is even considerable understanding of the cellular assembly of some immunoglobulins. Entire books can be written about the chemistry of antigens and of complement.

It is, in fact, the expanding wealth of knowledge in the field of immunochemistry that spawned this new series of books. So much information is available that the immunochemist finds it a struggle just to keep the major problems in his discipline in focus! Most of the articles in this and the forthcoming volumes will not be comprehensive reviews of the subject. It is the editors' intention to publish discussions of rather limited scope in areas of intensive research. The purposes of most articles will be to inform the reader of the present state of comprehension germane to that particular problem area, to describe the current research, and to offer suggestions for future inquiries. To obtain these goals, writers may choose to incorporate a considerable amount of their own data and to describe the problems as they view them. So that the series always remains true to its title, the editors and publisher alike shall endeavor to see that each work is published as rapidly as possible.

The subject matter of this series of books will be of interest, of course, to scientists actively engaged in research in the broad field of immunology. It is also intended that the articles will be suitable as supplemental reading material for graduate students. The explanation of immunology on a molecular level has developed prodigiously since the proposal of Ehrlich or even those of Burnet. It is the editors' desire that the articles in *Contemporary Topics in Immunochemistry* will, through reader apperception, set the proper direction for immunochemical research.

F. P. Inman

Athens, Georgia
November, 1971

Contents

The Relationship Between Antigenic Structure and Immune Specificity

E. Benjamini, R. J. Scibienski, and K. Thompson

I. Introduction	1
II. Antigens	2
A. Poly-γ-D-Glutamyl Capsule of *Bacillus anthracis*	2
B. Adrenocorticotropic Hormone (ACTH)	3
C. Angiotensin	4
D. Bradykinin	5
E. Gastrin	6
F. Glucagon	7
G. Ferredoxin	8
H. Insulin	10
I. Cytochrome *c*	12
J. Ribonuclease	15
K. Lysozyme	17
L. Myoglobin	20
M. Hemoglobin	22
N. Tobacco Mosaic Virus and Its Protein Subunit	25
III. Discussion	29
A. Reaction with Circulating Antibodies	29
B. Specificity of Cellular Immune Reactions	35
IV. Concluding Remarks	40
Appendix	40
References	43

Human Histocompatibility Antigens

Ralph A. Reisfeld and Barry D. Kahan

I. Introduction	51
II. Extraction of Histocompatibility Antigens	53

III. Solubilization and Purification of HL-A Antigens 57
    A. Antigen Source Material 57
    B. KCl Extraction Technique 58
    C. Purification with Polyacrylamide Gel Electrophoresis 59
IV. Biological Evaluation of Soluble HL-A Antigens 63
V. Chemical Characterization of HL-A Antigens 67
    A. Antigens Prepared from Spleen Cells 67
    B. HL-A Antigens from Lymphocytes in Culture 68
VI. Chemical Nature of HL-A Alloantigens 71
VII. Summary 74
Acknowledgment 74
References 74

Bacterial Flagellin as an Antigen and Immunogen

Christopher R. Parish and Gordon L. Ada

I. Introduction 77
II. Properties of Flagellin from *Salmonella* Organisms 78
    A. Physical and Chemical Properties 78
    B. Antigenic Properties 79
III. Degradation and Modification of Flagellin 79
    A. Antigenic Properties of Fragments Released from Flagellin 79
    B. Antigenic Properties of Chemically Modified Flagellin 80
IV. *In Vivo* Immunogenicity of Flagellin and Its Derivatives 81
    A. Immunogenicity of Flagellin and Polymerized Flagellin 81
    B. Immunogenicity of Fragmented Flagellin 82
    C. Immunogenicity of Chemically Modified Flagellin 83
V. *In Vivo* Localization Patterns of Flagellin and Its Derivatives 85
VI. *In Vitro* Behavior of Flagellin and Its Derivatives 86
    A. Induction of Immune Responses 86
    B. Reaction Between Antigen and Lymphocytes 88
VII. Antigen as a Regulator of Cell Behavior 89
VIII. Discussion and Conclusions 89
References 91

The Transfer of Immunity with Macrophage RNA

Joel W. Goodman

I. Introduction 93
II. The Heightened Immunogenicity of Antigens Associated
    with Macrophages 94
    A. Immune Responses of Fractionated Cell Populations 94
    B. Retention and Localization of Antigen by Macrophages 95

III. RNA–Antigen Complexes   96
  A. Assessment of the Requirement for Macrophages   96
  B. Assessment of the Existence of Antigen-Specific RNAs   100
  C. Assessment of the Requirement for an Enzyme   100
  D. Relationship Between the Capacity to Complex with RNA
     and Immunogenicity   100
  E. The Nature of the RNA–Antigen Chemical Bond   101
  F. Are RNA–Antigen Complexes Laboratory Artifacts?   102
IV. The Transfer of Cellular Immunity with RNA   103
V. Informational RNA   106
VI. Conclusions   108
Acknowledgment   109
References   109

Relationship of Events at the Lymphocyte Cell Surface to Gene Expression: Approaches to the Problem

Richard A. Lerner

I. Introduction   111
II. General Properties of Continuously Growing Cultured
    Human Lymphocytes   112
III. Control of Synthesis of Membrane-Associated Immunoglobulin . . .   113
  A. Quantitative Aspects   113
  B. Half-Disappearance of M-Ig in Logarithmically Growing Cells
     Treated with Inhibitors of Protein Synthesis   118
  C. Amount and Half-Disappearance of M-Ig and Cytoplasmic
     Ig in Synchronized "G_0" Cells   120
  D. Half-Disappearance of M-Ig and Cytoplasmic Ig in
     Logarithmically Growing Cells Treated With Actinomycin D . .   120
IV. Molecular Events During the Rest to Proliferation
    Transition in Lymphocytes   123
  A. Synchronization of Cultured Lymphocytes   123
  B. Phenotypic Expression in Stationary Phase ("G_0")
     and Logarithmically Growing Cells   124
  C. Polypeptide Synthesis in the "G_0" to G_1 Transition   126
  D. Perturbations Using Actinomycin D and Cyclic AMP   129
V. Studies on the "Linkage" Between the Plasma Membrane
    and Cellular Genes   130
  A. Nature of Cytoplasmic DNA   132
  B. Association of DNA with Cytoplasmic Membranes   132
  C. Synthesis of Membrane-Associated DNA in Synchronized Cells . .   136

D. Electron Microscopic Studies of Plasma Membrane-
        Associated DNA 138
VI. Summary and Prospects for the Future 141
Acknowledgment 142
References 142

The Antigen-Binding Sites of Immunoglobulins

R. R. Porter

I. Introduction 145
II. Size and General Features of the Antibody Combining Site. 146
III. Structural Studies. 149
IV. Sequence of the Variable Regions of Heavy and Light Chains 150
V. Affinity-Labeling Studies. 154
VI. Conclusion 160
References 163

γD Immunoglobulin

Hans L. Spiegelberg

I. Introduction 165
II. Isolation of γD 166
III. Structure of γD. 167
    A. Physical Properties 167
    B. Chemical Properties 169
IV. Biological Properties of γD 176
    A. γD Concentration in Body Fluids 176
    B. Antibody Activity 176
    C. Metabolic Properties 177
    D. Other Biological Activities 178
    E. γD Multiple Myeloma 178
V. Prospects 178
Acknowledgment 179
References 180

Index 181

The Relationship Between Antigenic Structure and Immune Specificity

E. Benjamini, R. J. Scibienski,* and K. Thompson†

Department of Medical Microbiology
School of Medicine
University of California
Davis, California

I. INTRODUCTION

There has built up over the years an accumulation of data dealing with the specificity of antigen-antibody interactions. Numerous studies, some utilizing admittedly ill-defined antigens, have made solid contributions to our knowledge concerning some of the structures and forces involved in immune interactions. These have been the subject of numerous reviews (Karush, 1962; Kaminski, 1965; Singer, 1965; Sela, 1966, 1967, 1969; Crumpton, 1967b; Kabat, 1966, 1968; Pressman and Grossberg, 1968; Goodman, 1969) and will therefore be given only cursory treatment herein. However, knowledge of the precise interrelationships involved in these interactions can best be derived from studies with antigens of defined structure. The present effort will be limited to coverage of selected protein antigens of known structure and/or sequence. It is not our intent to review the literature, but rather to focus our attention on several representative model proteins, in an attempt to extract from the diverse findings common denominators and to bring forth some unifying concepts dealing with various parameters of the immune response. We extend our apologies to those workers in the field whose contributions have not been included.

*Recipient of a Postdoctoral Fellowship from the National Institutes of Health (1-F02-AM 35714).
†Recipient of a Postdoctoral Fellowship from the Medical Research Council of Canada (100-2T-60).

1

We have chosen to divide our effort into two main sections. In the first, we shall present summaries of the extensive investigations which have been carried out with the antigens that have been selected. In the second section, we shall draw upon these findings, as well as upon pertinent supplementary information, in an attempt to construct a cohesive picture dealing with immune specificity at several levels of the immune mechanism. In this latter section, we have taken the liberty of referencing only the supplementary information which has not been covered in the preceding section. For the convenience of the reader, the sequences of the peptides and proteins dealt with in this chapter are presented in the Appendix. Where known, corresponding sequences from other species can be found in the *Atlas of Protein Sequence and Structure* (compiled by Dayhoff and Eck, 1969).

II. ANTIGENS

A. Poly-γ-D-Glutamyl Capsule of *Bacillus anthracis*

Immunization of animals with *Bacillus anthracis* gives rise to antibodies with specificity to the γ-D-glutamic acid capsular polypeptide. Although the isolated capsular polypeptide is of molecular weight 33,000, it is nonimmunogenic but is capable of acting as a hapten. Thus, when the peptide is electrostatically complexed with methylated bovine serum albumin, antibodies specific to the poly-γ-D-glutamyl peptide can be induced (summarized by Goodman, 1969). Moreover, while the pure poly-γ-D-glutamyl peptide is nonimmunogenic, it can induce unresponsiveness and is in fact a tolerogen in animals subsequently injected with the immunogenic form of the polypeptide (Roelants and Goodman, 1970).

Short polymers of γ-D-glutamic acid, prepared synthetically or by partial acid hydrolysis of the capsular polypeptide, have been shown to be reactive with antibodies (Nitecki and Goodman, 1966; Goodman *et al.,* 1968). With two sera, increments in antigenic activity were obtained up to the penta-γ-D-Glu peptide, which was as active as the hexamer; a third antiserum distinguished between the pentapeptide and the hexapeptide. These results suggest that the optimal size of the antigen involved in combining with rabbit antibodies to poly-γ-D-Glu is equivalent to that of a hexapeptide and indicate a size of approximately 36 by 12 by 7 Å (the size of hexaglutamic acid). In the same study, it was found that a mixture of larger peptides obtained from the partial acid hydrolysis of the capsular polypeptide was markedly more reactive than the hexa-γ-D-glutamic acid (Goodman *et al.,* 1968). It was concluded that antibodies were probably directed against a conformation of poly-γ-D-glutamyl peptide, as indicated by a correlation between the peptide length and acquisition of secondary structure in solution (Goodman, 1969).

B. Adrenocorticotropic Hormone (ACTH)

Adrenocorticotropic hormone is a straight-chain polypeptide of 39 amino acids. Of these, the *N*-terminal sequence 1-24 is common to ACTH from all the species investigated and is responsible for the hormonal activity of the molecule; species variations in sequences are found in the region 25-33.

Antibodies to the hormone (either natural or synthetic) are usually elicited by immunization with hormone or its fragments conjugated to carriers (Fischer *et al.*, 1965; Felber *et al.*, 1966; Felber and Micheli, 1967; Salvin and Liauw, 1967; Gelzer, 1968; Reichlin *et al.*, 1968*b*). Using antibodies so formed, studies revealed that a portion constituting the *C*-terminal part of the hormone (sequence 25-39) was antigenic and possessed a high capacity to inhibit the binding between ACTH and homologous antibodies (Fleischer *et al.*, 1965, 1966; Imura *et al.*, 1965; Gelzer, 1968). However, antibodies directed to sequence 1-24, capable of neutralizing the biological activity of the hormone, were demonstrated by several workers (Felber *et al.*, 1966; Felber and Micheli, 1967). Moreover, the free 1-24 peptide was found to be immunogenic in guinea pigs (Axelrod *et al.*, 1963; Salvin and Liauw, 1967). In addition, it was demonstrated that whereas the peptide 1-23 was not immunogenic, the same peptide, *N*-acetylated and with the ϵ-amino lysines formylated, was void of hormonal activity but was immunogenic (Axelrod *et al.*, 1963).

The antigenic specificity of peptide 1-24 has been investigated using antisera produced in rabbits following immunization with peptide 1-24 coupled to rabbit serum albumin (Gelzer, 1968). Compared to the inhibitory activity of peptide 1-24, the *C*-terminal 11-24 and 17-24 fragments were stronger inhibitors, whereas the *C*-terminal 20-24 peptide possessed very little activity. The conclusion was drawn that the combining site of the antibody was complementary to the octapeptide (17-24) or the tetradecapeptide (11-24). However, the finding that the substitution of the two arginine residues at positions 17 and 18 with ornithine did not drastically affect antigenicity may be indicative of a hexapeptide determinant composed of sequence 19-24. With regard to the stronger inhibitory activity of peptides 11-24 and 17-24 (compared with that of peptide 1-24), it was also reported (Imura *et al.*, 1965) that certain synthetic regions of ACTH were more antigenic than intact natural ACTH. This may possibly be because the antibodies formed in response to immunization with the conjugate recognize a conformation on the conjugate which cannot be assumed by peptide

1-24 in free solution, whereas the conformations of peptides 11-24 and 17-24 on the conjugate and in free solution may be similar.

Investigations into the role of the structure of peptide 1-24 in the induction of delayed hypersensitivity and the production of circulating antibodies in guinea pigs showed that the immunogenic part of the molecule (for induction of delayed hypersensitivity) consisted of residues 17-24. Furthermore, the 1-24 peptide was not only immunogenic but could also serve as a carrier for haptens: when it was used as carrier for azobenzoate, guinea pigs responded with delayed reactions specific to the carrier (i.e., peptide 1-24); sensitization with a conjugate of azoarsonate-peptide 1-24 induced delayed hypersensitivity with specificity toward the hapten (Salvin and Liauw, 1967).

C. Angiotensin

The action of renin on angiotensinogen, an inactive precursor molecule, releases a potent pressor decapeptide (angiotensin I) having the sequence Asp-Arg-Val-Tyr-Ile-His-Pro-Phe-His-Leu (horse angiotensin I), which may be further converted to angiotensin II (also an active peptide) following the enzymatic removal of the C-terminal dipeptide (Skeggs et al., 1957).

Probably because of its small size and susceptibility to enzymatic degradation, free angiotensin is usually not immunogenic. However, immunization with free angiotensin II amide (Asp-β-amide) physically adsorbed onto microparticles of carbon resulted in the production of antibodies to free angiotensin which could effectively neutralize its hormonal activity (Boyd and Peart, 1968). Angiotensin conjugated to carriers has been widely used for antibody production (Deodhar, 1960; Goodfriend et al., 1964; Haber et al., 1965; Goodfriend et al., 1966). Since the reaction of free angiotensin with antibodies is haptenic (with no precipitin formation and no ability to fix complement), its immunological activity (or that of its analogues) is commonly assessed by the inhibition of the reaction either between the polyvalent angiotensin-carrier conjugate or between isotopically labeled free angiotensin and antibodies.

Some studies on the relationship between structure and activity revealed the relative unimportance of the first N-terminal amino acid and led to the conclusion that the heptapeptide 2-8 or the hexapeptide 3-8 was responsible for antigenic activity. It was further concluded that the C-terminal phenylalanyl residue was of utmost importance for immunological specificity, while the residue at position 4 was not critical (Dietrich, 1967). On the other hand, assays of the immunological activity of several analogs indicated that the phenolic hydroxyl at position 4 and the two N-terminal acids were important for antigenicity. Moreover, disruption of the chain conformation by substituting the

D-isomer for the L-isomer at position 4 resulted in greatly reduced antigenic activity (Goodfriend *et al.,* 1966). The former conclusion is corroborated by the finding that angiotensin I did not react with antibodies produced against Val 5 angiotensin II (d'Auriac *et al.,* 1969). Also, it was shown that antibodies to synthetic angiotensin II reacted only slightly with antiotensin I (Catt and Coghlan, 1967). The importance of the *C*-terminal portion to the immunological activity was further indicated by studies with angiotensin analogues and fragments (Haber *et al.,* 1967). The discrepancies between the various conclusions may be accounted for by differences in the techniques used, and by the possible genetic differences of the animals used. In fact, it was shown that different rabbits immunized with Asn 1-Val 5 antiotensin II conjugate produced antibodies with different affinities to Asn 1-Val 5 angiotensin, Asp 1-Val 5 angiotensin, Asn 1-Ile 5 angiotensin, and the heptapeptide 2-8 (Hollemans *et al.,* 1968).

Although the activity of angiotensin can be neutralized by antibodies (Deodhar, 1960; Hedwall, 1968; Oken and Biber, 1968), there is evidence suggesting that the structural requirements for binding with antibodies differ from those necessary for binding with the biological receptors of the hormone: the hexapeptide 3-8, while possessing immunological activity, has only 2% of its pressor activity, and the monoiodo- and diiodo-Tyr 4 derivatives, which possess immunological activity, have only 25% and 1-4% of their pressor activity, respectively (Dietrich, 1967).

D. Bradykinin

Antibodies to bradykinin have been obtained by immunization of rabbits with a conjugate of bradykinin coupled to poly-L-lysine through the *N*-terminal portion of the peptide (Spragg *et al.,* 1966). Experiments on the relationship between structure and binding with antibodies, utilizing various analogues of the peptide, revealed some interesting phenomena regarding antigenic specificity (Haber *et al.,* 1967; Spragg *et al.,* 1967, 1968). The length of the peptide (nonapeptide) seemed to be of importance, since either adding an arginine to or removing an arginine from the *C*-terminus considerably affected the binding. However, replacing the *C*-terminal arginine with alanine or with D-arginine had little effect. It thus appears that the peptide bond between residues 8 and 9 contributed to binding, while the side chain of the amino acid at position 9 was not essential. In view of these findings, the authors raised the possibility that the peptide may exist in a cyclic form and that alterations in the length of the peptide adversely affected the stability of the ring. Additional investigations by the same workers indicated that the overall molecular charge did not contribute to binding, nor did the local charge at the *N*-terminal or the *C*-terminal arginine.

Furthermore, since the phenylalanine in position 8 or the serine in position 6 could be replaced by alanine without significantly affecting the binding, it was concluded that changes from hydrophilic to hydrophobic groups or vice versa had little effect.

Changes in the molecule which did cause a dramatic decrease in binding with antibradykinin were those which affected the backbone structure of the peptide: (a) substitution of the glycine in position 4 by alanine, leading to a constraint imposed on the native conformation due to the limitation of rotation by β carbon; (b) substitution of the proline in positions 2, 3, or 7, individually or simultaneously, by D-proline; (c) substitution of the proline in positions 2, 3, or 7 by alanine. These data point to the importance of each of the proline residues in the peptide (and especially of proline in position 3) and lead to the conclusion that a change in the overall shape of the peptide profoundly affects binding. This, coupled with the previously mentioned observation that peptide length is important, strongly suggests that the antigenic determinant comprises the entire peptide and that antibodies recognize the peptide in a preferred globular conformation (Haber *et al.,* 1967).

Whereas the relationship between structure and biological activity did not always correlate with the relationship between structure and antigenic activity, both were profoundly affected by changes in the backbone of the peptide. The lack of correlation may be due to differences between the conformation of the peptide in free solution (the form used to assay biological activity) and its conformation in poly-L-lysine (which is used for immunization).

E. Gastrin

Gastrin is a peptide hormone consisting of 17 amino acids. The natural gastrin molecule is termed "gastrin I." Synthetic gastrin, termed "gastrin II," differs from gastrin I by having the tyrosine residue at position 12 sulfonated (Anderson *et al.,* 1964). Gastrin molecules from all species investigated, including man, have very similar amino acid sequences and differ from each other by only one or two amino acids (Dayhoff and Eck, 1969). Moreover, the *C*-terminal pentapeptide amide of gastrin (Gly-Trp-Met-Asp-Phe-NH$_2$) is present in the hormone of all the animal species investigated and is responsible for the physiological activity of the peptide (Gregory, 1966).

Antibodies to gastrin have been evoked in rabbits by immunization with synthetic human gastrin I (sequence 2-17) conjugated to a protein carrier. These antibodies were demonstrated to bind with the radioiodinated homologous antigen and with gastrin molecules obtained from several other mammalian species (porcine, canine, ovine) (McGuigan 1968*a,b*). These cross-reactions are not surprising in view of the findings that the antibodies were directed against the *C*-terminal tetrapeptide amide, which is shared by all gastrins studied. The antibodies inactivate the physiological activity of gastrin. Moreover, caerulein and cholecystokinin pancreozymin (both physiologically active peptides), which possess the same C-terminal pentapeptide amide as gastrin (Gregory, 1966; Mutt and Jorpes, 1967; Anastasi *et al.*, 1968), were reactive with these antibodies (McGuigan, 1969*a,b*).

The antigenic C-terminal tetrapeptide amide, conjugated to a protein carrier, was used for immunization of rabbits. The resulting antibodies, which neutralized hormonal activity, reacted with synthetic human gastrin, with naturally occurring porcine gastrin, with the C-terminal pentapeptide of gastrin, as well as with the C-terminal tetrapeptide (McGuigan, 1967). These antibodies reacted equally well with the C-terminal tetrapeptide as with the entire gastrin molecule, indicating that the tetrapeptide resided in an exposed position on the surface of the intact hormone and, furthermore, that the conformation of the gastrin molecule did not affect the spatial properties of the tetrapeptide. Interestingly, antibodies produced by immunization with synthetic human gastrin (2-17) conjugate reacted with the hormone cholecystokinin pancreozymin, but substantially less well than they did with gastrin. On the other hand, antibodies elicited in response to immunization with the conjugate of the C-terminal tetrapeptide amide reacted equally well with intact gastrin and with cholecystokinin pancreozymin (McGuigan, 1968*b*). This, together with the finding that the C-terminal tetrapeptide of gastrin reacted considerably less well than did gastrin with antibodies to synthetic human gastrin (2-17) conjugate, implies that determinants other than just the C-terminal tetrapeptide amide may be involved. However, the fact that peptide 1-13 did not exhibit demonstrable binding argues against such determinants. Apparently, peptide 1-13 participates in the antigen-antibody interaction, but whether or not this participation is specific and depends upon the intact sequence of peptide 1-13 is not clear. On the other hand, this peptide may enhance the binding of the C-terminal tetrapeptide amide in a nonspecific manner. However, the reduced reactivity of cholecystokinin pancreozymin with antibodies to gastrin conjugate (as compared to the binding with the homologous peptide) points to some degree of specificity exhibited by peptide 1-13.

F. Glucagon

The physiologically active peptide glucagon is a single-chain molecule composed of 29 amino acids. In spite of its relatively low molecular weight, glucagon is immunogenic and exhibits extensive immunological cross-reactivity. Thus rabbit antibodies produced in response to immunization with ox or pig glucagon reacted with dog and human glucagon (Unger *et al.*, 1961). However, little or no reaction could be demonstrated between anti-ox glucagon and glucagon obtained from the sculpin fish (*Cottus scorpius*) (reviewed by Wilson, 1967).

Antibodies obtained from guinea pigs immunized with glucagon emulsified in complete Freund's adjuvant exhibited specificity primarily for the amino terminal region of the molecule. Of the region which was demonstrated to bind with antiglucagon, the peptides Ser 2-Ser 16, Gln 3-Ser 16, Gly 4-Ser 16, and Thr 5-Ser 16 were active (although considerably less so than intact glucagon), whereas peptide Phe 6-Ser 16 was inactive. The *N*-terminal region (His 1-Arg 17 or His 1-Arg 18) and the *C*-terminal region (Arg 18-Thr 29 or Ala 19-Thr 29) elicited delayed skin reactions in glucagon-immunized guinea pigs and were able to inhibit the migration of peritoneal exudate cells derived from such guinea pigs. The *C*-terminal peptide stimulated DNA synthesis by splenic lymphocytes derived from glucagon-immunized guinea pigs to almost the same degree as did glucagon, whereas the *N*-terminal peptide was inactive in this respect. A mixture of the *N*-terminal peptide and the *C*-terminal peptide did not increase the stimulatory activity of the latter (Senyk *et al.*, 1971*a,b*). Regarding the stimulatory activity of the *C*-terminal peptide, it was observed that Ala 19 in the undecapeptide Ala 19-Thr 29 was essential, since the peptide Gln 20-Thr 29 was unable to stimulate. Moreover, the addition of arginine to the *N*-terminal of the above undecapeptide resulted in a dodecapeptide with markedly decreased stimulatory activity on cells from most, but not all, animals. On one hand, the alteration in activity may be due to the additional charge conferred by arginine or to an alteration in the conformation of the peptide, which would imply that the conformation of the undecapeptide in the glucagon molecule is more similar to its conformation in free solution than to its conformation in the dodecapeptide. On the other hand, perhaps the free undecapeptide can readily assume a conformation complementary to the cell receptor, whereas the *N*-terminal addition of arginine imposes a constraint.

G. Ferredoxin

Ferredoxin of *Clostridium pasteurianum* and its performic acid-oxidized form elicited antibody formation in rabbits following immunization with the proteins in Freund's complete adjuvant. It was found that when native ferredoxin was used as immunogen the rabbit antibodies reacted much better with ferredoxin from which the iron was removed (oxidized ferredoxin). This phenomenon implies that prior to the initiation of the immune response the iron may have been removed *in vivo*. The reaction of these antibodies with various derivatives of ferredoxin showed that the eight cysteine residues (non-disulfide-linked in the native molecule) played only a minor role in antigenicity (Nitz *et al.*, 1969). Moreover, it was concluded that the regions between Cys 8 and Cys 18, and Cys 37 and Cys 47, play little role, if any, in the antigenicity of *C.*

pasteurianum ferredoxin. Accordingly, the search for antigenic determinants was focused on other areas of the molecule (Mitchell *et al.,* 1970). The peptides comprising residues Lys 3-Phe 30 and Val 31-Glu 55 were able to inhibit the reaction between performic acid-oxidized ferredoxin and homologous antiserum. On the other hand, the regions between Cys 8-Cys 18 and Cys 37-Cys 47 were apparently not antigenic (by the reasoning given earlier). The authors therefore concluded that peptides Lys 3-Ser 7, Pro 19-Phe 30, Val 31-Thr 36, and Pro 48-Glu 55 might contain antigenic determinants. Consequently, peptide Ala 1-Ser 7 (Ala-Tyr-Lys-Ile-Ala-Asp-Ser), peptide Val 31-Thr 36 (Val-Ile-Asp-Ala-Asp-Thr), and peptide Pro 48-Glu 55 (Pro-Val-Gly-Ala-Pro-Val-Gln-Glu) were synthesized and assayed for immunological activity. Of these, the *N*-terminal heptapeptide and the *C*-terminal octapeptide exhibited significant activity (Mitchell *et al.,* 1970; Kelly and Gerwing Levy, 1971). Ferredoxin of *Clostridium butyricum,* which differs from that of *C. pasteurianum* at nine positions, showed strong reactivity with antiserum to *C. pasteurianum* ferredoxin and its performic acid-oxidized derivative. It is therefore interesting that a synthetic peptide representing the *C*-terminal octapeptide of *C. butyricum* ferredoxin (differing from the analogous peptide of *C. pasteurianum* ferredoxin in having Asn instead of Val in position 53) reacted only slightly with antibodies to *C. pasteurianum* ferredoxin but reacted with antibodies to performic acid-oxidized ferredoxin of *C. pasteurianum* as well as did the homologous peptide.

Further investigations (Christensen *et al.,* 1971) showed that significant antigenic activity of the *C*-terminal octapeptide of ferredoxin resided in the *C*-terminal tetrapeptide, with marginal activity exhibited even by the *C*-terminal tripeptide. The *C*-terminal pentapeptide and the entire *C*-terminal octapeptide exhibited equal binding with antibodies to performic acid-oxidized ferredoxin. These findings indicate that the antigenic determinant could be a tetrapeptide and that the additional *N*-terminal amino acids greatly enhance binding.

More recently, the same workers conjugated the *N*-terminal heptapeptide, the *C*-terminal tetrapeptide, or both peptides to a protein carrier. Each conjugate was shown to be reactive with antibodies to oxidized ferredoxin. Moreover, antibodies induced by immunization with the conjugates exhibited reactivity not only with the homologous peptides but also with oxidized ferredoxin (Teather and Gerwing Levy, 1971). In guinea pigs sensitized with performic acid-oxidized ferredoxin, the free peptides and their conjugates were able to elicit delayed skin reaction and to inhibit the migration of peritoneal exudate cells. However, neither the free peptides nor the single peptide conjugates were able to stimulate DNA synthesis in lymphocytes derived from oxidized ferredoxin-sensitized animals. In contrast, the conjugate containing both the *C*- and *N*-terminal peptides was extremely stimulatory—in fact, more so than the antigen used for immunization (Waterfield *et al.,* 1971).

H. Insulin

Insulin is a protein of relatively low molecular weight composed of 51 amino acids. The molecule consists of two polypeptide chains, the A chain having 21 amino acid residues and the B chain having 30 residues. The chains are linked to each other by two disulfide bridges (Cys_A 7-Cys_B 7 and Cys_A 20-Cys_B 19). In addition, there is one intrachain disulfide bridge on the A chain between residues Cys_A 6 and Cys_A 11. The complete sequences of insulins from several sources have been determined (compiled by Dayhoff and Eck, 1969).

The immunology of insulin has been extensively covered in several reviews (Arquilla, 1962; Prout, 1962; Pope, 1966; Schwick, 1966; Wilson, 1967). Many mammalian insulins have been found to be immunologically cross-reactive, the cross-reactions extending to the neutralization of hormonal activity. However, the insulins of the coypu (Davidson *et al.*, 1968), the capybara (Davidson *et al.*, 1969), and the guinea pig (Moloney and Coval, 1955; Kitagawa *et al.*, 1960; Davidson and Haist, 1965) have been shown to be non-neutralizable by antibodies to beef insulin. Also, capybara insulin has been shown to be non-neutralizable by guinea pig antibodies to chicken and to cod insulins (Davidson *et al.*, 1969).

Human antibodies to beef insulin have been shown to be reactive with endogenous human insulin, although the interaction between beef insulin and these antibodies was stronger (Grodsky, 1965). Antibodies of the IgE class, produced in humans in response to beef-pork therapeutic insulin, were shown to react with sheep insulin and with extracted autologous insulin but did not react with isolated A or B chains (Patterson *et al.*, 1969). In another report (Berson and Yalow, 1959), it was shown that human anti-beef-pork therapeutic insulin was more reactive with beef and sheep insulins than with pork or horse insulins, while human insulin was the least reactive. Since this order of reactivity parallels the known differences in amino acid sequence of these proteins, it was postulated that the region comprising residues 8-10 of the A chain (where the differences occur) may be involved in antigenicity. No differences were found between the binding of iodinated and native insulin by this serum, nor did the antibodies distinguish between native and desamido insulin; these antibodies could not be shown to bind iodinated A or B chains.

Since studies with human antibodies to pork insulin (which differs from human insulin only in the *C*-terminal residue of the B chain) showed that the antibodies reacted with a derivative of pork insulin from which this amino acid was removed, it may be concluded that the antibodies reacted with sequences of pork insulin which are common to human insulin (Berson and Yalow, 1963).

Studies on the reactivity of human anti-beef insulin with various chemically modified derivatives of the molecule showed that dinitrophenylated (lysines only), acetylated (85% of the available amino groups), desalanylated, sulfanylated, and methylesterified (at six out of 12 groups) insulins were almost identical to native insulin in their reaction with antibodies. Methylesterified insulin (with 11 out of 12 groups modified) and oxidized insulin were only poorly reactive (Grodsky *et al.*, 1959). Beef insulin which had been reacted with fluorescein isothiocyanate was less reactive with rabbit antibodies than was native insulin was (Halikis and Arquilla, 1961), although a later report (Arquilla *et al.*, 1966) indicated that the reactivity of this derivative with some antibodies was unchanged. A reduction in the antigenic activity of insulin which had been fluoresceinated at 0.9 groups per molecule has also been reported (Brunfeldt *et al.*, 1968). In addition, it was shown that iodination reduced the antigenic activity of insulin, the loss of activity being proportional to the degree of iodination, although some activity remained even at ten iodine atoms per molecule. Other studies on the effect of iodination on the antigenic activity of insulin showed that less than one atom per molecule resulted in reduced activity. These findings, coupled with earlier findings on iodinated derivatives of insulin (De Zoeten and De Bruin, 1961; De Zoeten and Havinga, 1961), indicated that the *C*-terminal region of the A chain of insulin plays an important role in the antigenicity and biological activity of the molecule (Arquilla *et al.*, 1966, 1968).

Several derivatives of insulin (iodinated, fluoresceinated, and desoctapeptide insulin) have been used to study the response of strain 2 and strain 13 guinea pigs to insulin (Arquilla *et al.*, 1969). It was found that strain 2 guinea pigs produced antibodies which were predominantly directed against areas involving the *N*-termini of both the A and B chains, while strain 13 guinea pigs produced antibodies directed predominantly against a region involving the *C*-termini of both chains.

Studies with the isolated A and B chains of insulin have demonstrated that whereas anti-A chain reacted only with the homologous antigen, antibodies to the B chain reacted with both B chain and intact insulin (Yagi *et al.*, 1965; Varandini, 1967). In addition, anti-insulin was found to react with isolated B chains (Yagi *et al.*, 1965). Subsequently, it was reported that the B-chain fragments, B1-22, B1-16, and B1-10 exhibited reactivity with several antisera to B chain, while fragment B23-30 exhibited activity with fewer antisera. The fragments B11-16 and B17-25 were inactive with all the antisera tested (Yagi *et al.*, 1967). It has also been reported that human antibodies to beef insulin were reactive with isolated B chains but not with A chains. However, when two sheep antisera to sheep insulin and a pig antiserum to pig insulin were studied, it was found that A chains had slight activity, the B chains had intermediate activity, and intact insulin was the most active (Kerp *et al.*, 1969). These latter studies, however, were carried out by measuring the displacement of iodinated insulin from antibodies. Therefore, the results with A chain should be carefully interpreted in light of the previously mentioned studies which indicated that A-chain antigenicity is readily affected by iodination (Arquilla *et al.*, 1966, 1968).

Studies on the reactivity of various B-chain derivatives with anti-insulin antibodies indicated that the synthetic disulfide-linked dimers $B(1-8)_2$ and $B(17-30)_2$ were reactive with antibodies but that $B(2-8)_2$, $A(13-21)_2$, and $A(10-21)_2$ were not (Kerp *et al.*, 1968; Wilson *et al.*, 1967). In addition, it was reported that B1-8, B9-30, and B24-30 were reactive with anti-B chain, whereas $B(2-8)_2$, B9-14, and $B(17-23)_2$ were not. Also, studies on the reaction of A-chain derivatives with anti-A demonstrated that A1-9 and A10-21 were reactive and that intact A chain could inhibit the insulin-anti-insulin reaction (Wilson *et al.*, 1967).

Investigations on the cross-reaction between the chains of beef, sheep, and pork insulins revealed that the A chains of beef and pork insulins (which differ at positions 8, 9, and 10) reacted similarly (but not quite identically) with anti-beef A chain, while sheep A chain was not so reactive. The three B chains (which are all identical in sequence) could not be distinguished by antibodies (Varandini, 1967).

Studies on the antigenic activity of sulfonated, oxidized, or reduced A and B chains indicated that antibodies to S-sulfonated A chain were reactive with the homologous antigen only; antibodies to oxidized or reduced A chains reacted with all three derivatives, as well as with intact insulin and with proinsulin. In addition, antibodies to any of the B-chain derivatives (sulfonated, oxidized, or reduced) reacted with all three derivatives, with insulin, and with proinsulin (Touber *et al.*, 1970).

An interesting investigation on the immunogenicity of insulin in humans showed that antibodies produced by two diabetic patients in response to treatment with beef insulin also reacted with human insulin. However, these patients ceased to produce these antibodies when the administration of beef insulin was discontinued, suggesting that human insulin is nonimmunogenic (in humans) but is recognized by antibodies to beef insulin (Grodsky, 1965). Another study showed that the removal of the C-terminal residue from the B chains of human, pork, and rabbit insulins resulted in identical molecules which were immunogenic in all three species (Lockwood and Prout, 1962).

I. Cytochrome *c*

Cytochrome *c* is a protein of molecular weight 12,400, the primary structure of which is known for many species (Dayhoff and Eck, 1969). A representative sequence of cytochrome *c* is given in the Appendix. Immunological studies with cytochromes *c* from many species have been used to demonstrate the correlation between primary structure and antigenic activity (Margoliash *et al.*, 1967*a,b*; Watanabe *et al.*, 1967).

Cytochromes are usually only weakly immunogenic in rabbits. This may be due to their small size and/or the retention of much of the primary structure throughout the phylogenetic scale. However, a high degree of immunogenicity can be achieved when cytochromes c are conjugated to carriers or are polymerized (by the use of ethanol or glutaraldehyde) (Reichlin et al., 1968a). Human cytochrome c was found to be the only native protein monomer which is a good immunogen. When tested with the monomeric homologous antigen, precipitating antibodies were found in antisera to human, monkey, and horse cytochromes c but not in antisera to tuna and turkey cytochromes c (Stavitsky, 1954; Margoliash et al., 1970). However, all antibodies did form precipitates and fixed complement with the polymerized homologous antigen. Excess of the monomeric forms can generally completely inhibit (highly efficiently) the precipitin and complement fixation reactions of polymers with the antisera, indicating the absence of polymer-specific determinants (Margoliash et al., 1967a,b; Reichlin et al., 1970).

Although antibodies to the cytochromes c of human, monkey, horse, tuna, kangaroo, and turkey reacted to varying extent with the cytochromes c of 25 species, species-specific determinants have been demonstrated. The non-species-specific (cross-reacting) antibody populations were in general directed against antigenic determinants that were immunologically similar (but not necessarily chemically similar) in all cross-reacting proteins, since there was no additivity with antibodies when two cross-reacting proteins were combined in the assay. Proteins having identical primary structures invariably turned out to be antigenically identical, but the reverse was not always true. Some of the sera could not distinguish between cytochromes c with different amino acid sequences, possibly because the differences were not present in antigenically active regions or were not in regions affecting conformation of the molecule (Margoliash et al., 1967a,b, 1970).

Antibodies from rabbits immunized with several cytochromes c reacted with rabbit cytochrome c. In addition, rabbits immunized with rabbit cytochrome c conjugated to bovine γ globulin produced antibodies capable of binding autologous cytochrome c (Nisonoff et al., 1969). These results indicate that with appropriate immunization an animal can produce antibodies to some autologous proteins.

Circular dichroism and optical rotary dispersion analyses could not reveal structural differences in several cytochromes c (Margoliash and Schejter, 1966). Therefore, it is reasonable to assume that the serological distinction of various cytochromes c is related to differences in their amino acid sequences. Studies on two cytochromes c which differ in only one amino acid residue afforded the localization of certain regions within the sequence which contribute to antigenicity. In one study, human, horse, and donkey cytochromes c, which differ at position 47, were compared. Horse cytochrome c has a threonine residue, while human and donkey cytochromes have serine residues at this position. Rabbit anti-horse cytochrome c could not distinguish between horse and donkey cytochromes c, whereas rabbit anti-human cytochrome c was capable of distinguishing between these two cytochromes. It was concluded that the side chains of serine and of threonine affected the geometry of the determinant but did not actually participate in binding. Thus antibodies able to conform to a threonine-containing site (one methyl group larger than serine) could accommodate the smaller serine residue, but not vice versa (Margoliash et al., 1967a,b). Alternatively, this area may have been immunosilent when horse cytochrome c was used as immunogen in the rabbit, whereas the antigenicity of the area was expressed when the human protein was used as immunogen. In another study, human and monkey (Macaca mulatta) cytochromes c, differing at position 58 (human having an isoleucine residue and monkey having a threonine residue), were compared using rabbit antibodies. Antiserum to the human protein was shown to contain 40% more antibodies reactive with the homologous than with the heterologous antigen. In contrast, antibodies to monkey cytochrome c reacted identically with monkey and human cytochromes c. Therefore, the presence of the hydrophobic isoleucine correlated with the occurrence of an immunogenic determinant, whereas in the absence of isoleucine this region was immunogenically silent. However, antiserum to horse cytochrome c (with threonine at position 58) was more reactive with monkey than with human cytochrome c. Absorption of anti-human cytochrome c with an excess of monkey cytochrome c yielded a serum containing antibodies to a single determinant which was termed the "isoleucine site." Only those cytochromes c which have an isoleucyl residue at this position (human and kangaroo) reacted with these specific antibodies. Those antibodies in antisera to human cytochrome c which reacted with both the human and the monkey proteins did so equally well, indicating that the identity in amino acid sequence outside the "isoleucine site" correlates with identity in antigenic structure. Antibodies directed to the isoleucine site accounted for 30-40% of the total rabbit antibodies produced to human cytochrome c (Nisonoff et al., 1970). Studies with Fab fragments from anti-human and anti-monkey cytochromes c substantiated the single antigenic difference between the two molecules: with Fab from antihuman serum, each mole of monkey cytochrome c was shown to bind 3 mole of Fab, whereas each mole of human cytochrome c was shown to bind 4 mole of Fab. As expected, no differences in binding with Fab from antibodies to monkey cytochrome c were observed: both monkey and human cytochromes c bound 3 mole of Fab per mole protein (Williams and Reichlin, 1968). Comparisons of the reactivity of cytochromes c which differ by more than one residue indicated that the molecules contained a limited number of antigenic determinants. Investigations of the kinetics and the stoichiometry of the reaction between antibodies and cytochrome c also led to the conclusion that the number of antigenic sites on the molecule is limited (Noble et al., 1969).

   An interesting finding was that ferricytochromes c (oxidized forms) from horse, okapi, cow, and kangaroo fixed more complement (10-15%) with some antisera to horse cytochrome c than did the corresponding ferrocytochromes c (reduced forms) (Margoliash et al., 1967a,b), confirming conclusions from chemical studies that there is a conformational difference between the ferro- and ferri-forms of the molecule.

J. Ribonuclease

Bovine pancreatic ribonuclease (RNase-A) is a low molecular weight (12,000) protein which has been extensively characterized (Anfinsen, 1962; Avey et al., 1962; Crestfield et al., 1962; Gross and Witkop, 1962; Kartha et al., 1967; Scheraga, 1967). The amino acid sequence (see Appendix) and three-dimensional structure of the enzyme have been elucidated (Anfinsen, 1962, and Kartha et al., 1967, respectively). Another form of the enzyme, RNase-B, has been isolated from natural sources, and appears to differ from RNase-A in the replacement of one amide group with a carboxyl group (Tanford and Hauenstein, 1965). Treatment of RNase-A with the proteolytic enzyme subtilisin results in cleavage of the peptide bond between residues 20 and 21, leaving the molecule unaltered in other respects. This derivative, termed "RNase-S," is enzymatically active, but activity is lost when the N-terminal eicosapeptide (S-peptide) is removed resulting in a molecule termed the S-protein (Richards and Vithayathil, 1959).

Rabbit antibodies to RNase-A could not distinguish between the A and B forms of the enzyme (Brown et al., 1959a). Also, early antibodies to bovine RNase-A did not react with porcine RNase-A, although they were highly reactive with ovine RNase-A. However, with antisera from later bleedings not only was the cross-reaction between bovine and ovine RNase-A increased, but also porcine RNase-A was cross-reactive (Brown et al., 1960).

Treatment of RNase-A with urea or by boiling had no effect on its binding with antibodies, whereas alkaline denaturation reduced its antigenic activity somewhat and performic acid oxidation completely eliminated it (Brown et al., 1959a,b). However, some binding between oxidized RNase-A and antibodies to the native form was later demonstrated (Brown et al., 1967). The reverse reaction (binding between native RNase-A and anti-oxidized RNase-A) could not be detected (Brown, 1963). It may be significant that those antibodies to native RNase-A which were reactive with oxidized RNase-A were induced with the aid of Freund's complete adjuvant or alum-precipitated antigen. Thus the presence of denatured forms in the immunogens cannot be ruled out. Partially reduced RNase-A retained some reactivity with antibodies to the native form (Brown et al., 1959b). Moreover, it was shown in a later study that the completely reduced and randomly reoxidized RNase-A retained some reactivity with antibodies to the native form, again induced with complete Freund's adjuvant. This was attributed to determinants of a size which would be unaffected by conformational alterations (Mills and Haber, 1963).

Studies of the reaction between antibodies to native RNase-A and RNase-S or RNase-S-protein demonstrated that RNase-S had diminished activity (compared to RNase-A) and that the reactivity of the S-protein with antibodies was even lower; the RNase-S-peptide was inactive (Singer and Richards, 1959; Merigan and Potts, 1966). It was also shown that RNase-A did not react completely with antibodies to RNase-S and that denaturation of RNase-S or of the S-protein resulted in a loss of reactivity with these antibodies. These findings may be significant when considered in conjunction with the three-dimensional models of RNase-A (Kartha *et al.*, 1967) and RNase-S (Wyckoff *et al.*, 1967): the major difference between these two forms of the protein is that in RNase-S the *C*-terminus of the S-peptide (residue 20) and the *N*-terminus of the S-protein (residue 21) have moved out and away from each other to assume random coil configurations. It is thus probable that at least one antigenic determinant of RNase-A is located in the area of the peptide bond between residues 20 and 21. It is improbable that the determinant involves much of the S-peptide in light of the inability of the latter to inhibit the reations between RNase-A and homologous antibodies.

Reports of investigations with other derivatives of RNase-A have alluded to the possible location of two other determinants. It has been reported that *a*-deamination of lysine 1 resulted in a loss of some antigenic activity (Brown *et al.*, 1959b; Van Vunakis *et al.*, 1960). It was further reported that dinitrophenylation of the *a*-amino resulted in diminished reactivity, whereas guanidation of the ε-amino of the same *N*-terminal lysine had no effect on reactivity with antibodies. Also, removal of the *N*-terminal lysine had a dramatic effect on reactivity with antibodies, and the removal of the *N*-terminal dipeptide (lysyl–glutamic) had an even greater effect (Brown *et al.*, 1959b). It has been suggested that the glutamic acid at position 2 was involved in the binding of the S-peptide to S-protein (Finn and Hofmann, 1965). It is thus possible that the glutamic acid may be directly involved in binding with antibodies or may contribute to the conformational integrity of the *N*-terminal portion of the chain, or both. Another possible determinant of RNase-A has been suggested by studies with the carboxypeptidase-degraded molecule (Brown *et al.*, 1959b). Removal of the *C*-terminal valine had no effect on antigenic activity, whereas removal of the *C*-terminal tetrapeptide (Asp–Ala–Ser–Val) diminished the reactivity with antibodies. Thus it is possible that the *C*-terminal region also constitutes an antigenic determinant.

The reactivity of antibodies to oxidized RNase-A with several other denatured forms of the enzyme indicated that the cysteine residues (but not the methionine residues) were important for antigenicity (May and Brown, 1968). Also, the reaction between oxidized RNase-A and homologous antibodies has been shown to be inhibited by peptides 38-61 and 105-124 (containing 13 out of 17 nonpolar residues) from the oxidized molecule. Removal of amino acids 38 and 39 from the peptide 38-61 greatly diminished but did not eliminate its reactivity (Brown, 1962). Acetylation of peptide 38-61 at residues 41 and 61 has been shown to reduce its inhibitory activity by 80% (Brown, 1963). Since dinitrophenylation of the ε-amino of lysine 41 of intact oxidized RNase did not affect its reactivity with antibodies (Brown et al., 1967), it is probable that in the acetylated peptide it is the modification of residue 61 which results in the lowered reactivity.

Studies with antibodies which neutralize the enzymatic activity of RNase-A have yielded some interesting results. It has been shown that such neutralizing antibodies were more effective in the presence of large substrates and that higher-affinity antibodies were effective inhibitors of enzymatic activity. It was concluded that neutralization was a result of steric hindrance (Branster and Cinader, 1961). It was later demonstrated that neutralizing and non-neutralizing antibodies could compete with each other to some extent, possibly due to their reacting with overlapping sites on the RNase-A molecule (Cinader and Lafferty, 1963). It is also possible that each type of antibody induces a conformational alteration which results in a loss of structures recognized by the other type of antibodies. Such alterations have been shown to exist (Cinader, 1967). In fact, one rabbit consistently produced antibodies which activated the enzyme. The activation was found to be more pronounced with low molecular weight than with high molecular weight substrates. Moreover, with high molecular weight substrates the activation reached a peak and then fell off, suggesting that neutralizing antibodies were still able to bind to the activated enzyme. Additionally, it was found that the presence of substrate in the active site of the enzyme prevented the binding of the activating antibodies. Since the binding of the substrate is known to induce a conformational change (Sela et al., 1957), and in view of the fact that neutralizing antibodies could bind with the activated enzyme, it was concluded that the activating antibodies were binding to sites distant from the active site but were holding the molecule in a conformation more favorable for enzyme action.

More recently, an apparently different type of activating antibody (active with low molecular weight substrates) was found in a chromotographic fraction of neutralizing antibodies (Suzuki et al., 1969). It was subsequently shown that the activating antibodies could compete with neutralizing antibodies (Pelichova et al., 1970), suggesting that the former were directed against a site(s) adjacent to the active site of the enzyme.

K. Lysozyme

Lysozyme is a single-chain polypeptide of 129 amino acids, the sequence of which is shown in the Appendix. In addition to its primary structure, the three-dimensional configuration of the enzyme has been established by X-ray crystallography (Blake *et al.*, 1965).

It was demonstrated that two large peptides of lysozyme have antigenic activity (Shinka *et al.*, 1962, 1967*a,b;* Fujio *et al.*, 1959, 1962, 1968*a,b*). The activity of these peptides was demonstrated by their ability to inhibit the reaction between lysozyme and rabbit antilysozyme and by the direct binding of ^{14}C-labeled *N*-acetyl peptides with antibodies. These two peptides encompass the regions comprising amino acids 57-107 (area I) and 1-27 and 122-129 (area II). Disulfide bonds between cysteines 6 and 127, 64 and 80, and 76 and 94 were intact in these fragments. Of the total precipitable antibodies contained in a pool of antilysozyme serum, 47% was directed to area II and 12% to area I. Since the presence of one peptide did not influence the binding of the other with antilysozyme, it was assumed that the antiserum contained at least two different antibody populations (Fujio *et al.*, 1968*b*). Peptic digestion of lysozyme yielded an immunologically active peptide consisting of residues 64–83 and 91-107 linked by two disulfide bridges at positions 64 and 80, and 76 and 94 (Canfield and Liu, 1968). More recently, it was shown that within the large peptide (residues 57-107), the area confined by residues 60-83, with a disulfide bridge between Cys 64 and Cys 80 (the "loop" peptide), is an antigenic determinant of native lysozyme. Antibodies to the "loop" peptide have been obtained by adsorption of antilysozyme with a "loop" peptide immunoadsorbent. Moreover, immunization with loop peptide conjugated to poly-D,L-alanyl-poly-L-lysine produced antibodies which combined with native lysozyme as well as with the isolated loop peptide (Arnon, 1968; Arnon and Sela, 1969; Maron *et al.*, 1971). The reduced and alkylated "loop" peptide did not bind with antilysozyme serum but did bind appreciably with antiloop peptide sera, indicating that in antilysozyme serum the antibodies were directed to conformation-dependent determinant(s) whereas the antiloop serum may have contained some antibodies which recognized sequential determinants.

It has been shown by workers using different animal species that the completely reduced and alkylated lysozyme (CM-lysozyme) and the native form of the protein show minimal, if any, humoral cross-reactivity (Shinka *et al.*, 1967*a;* Gerwing and Thompson, 1968; Young and Leung, 1970). Furthermore, several peptides obtained form CM-lysozyme did not inhibit the reaction between native hen egg-white lysozyme and its homologous antiserum but did appreciably inhibit the reaction between CM-lysozyme and its homologous antibodies. Thus determinants of the native molecule may involve areas encompassing more than just linear sequences, or areas dependent on the intact three-dimensional structure of the molecule not present in CM-lysozyme. The CM-lysozyme determinate areas consist of the sequences 1-12 and 106-129 (part of lysozyme area II) (Young and Leung, 1970) and the sequence 74-96 (part of area I) (Gerwing and Thompson, 1968). Although the antigenic areas of lysozyme and of CM-lysozyme appear to overlap, further work is required in order to determine whether CM-lysozyme and lysozyme share common determinants. An interesting finding was the fact that the N-terminal peptide (residues 1-12) could appreciably inhibit the binding of the C-terminal peptide with anti-CM-lysozyme, indicating that in the denatured CM-lysozyme molecule the N- and the C-terminal regions may be in close proximity to one another (Young and Leung, 1970).

A cyanogen bromide derivative of hen egg-white lysozyme which had the disulfide bonds intact but which was cleaved on the carboxyl side of the methionine residues at positions 12 and 105 was still capable of inhibiting the reaction between lysozyme and antilysozyme by as much as 70%, indicating that much of the antigenic structure of the native molecule was retained (Bonavida *et al.*, 1969). N-bromosuccinic acid treatment of lysozyme (in which the tryptophan at position 62 or 63 was modified) resulted in a derivative which could still completely inhibit the reaction between lysozyme and antilysozyme. However, extensive modification (all six tryptophan residues) with 2-nitrophenylsulfonyl chloride completely destroyed the ability of the derivative to react with antinative lysozyme. This derivative was also found to be very weakly immunogenic compared to native lysozyme. In contrast to the tryptophan residues, when all seven lysine residues were succinylated the derivative still inhibited as much as 42% of the reaction between lysozyme and antilysozyme (Bonavida, 1968).

Whereas the C-terminal CNBr peptide of CM-lysozyme was unable to inhibit the reaction between native lysozyme and antinative lysozyme, it could inhibit the reaction between native lysozyme and antibodies raised in rabbits in response to immunization with CNBr-treated lysozyme. These observations led to the conclusion that there are determinants on hen egg-white lysozyme, termed "heterotopes," to which antibodies are not directed when the native molecule is used for sensitization but which are expressed when immunization is performed with a derivative, or with lysozyme from another species (Bonavida, 1968; Bonavida and Sercarz, 1971).

Extensive immunochemical comparisons have been performed on lysozymes from different origins, some of which led to the immunological prediction of sequence differences among the proteins (Arnheim and Wilson, 1967; Arnheim *et al.*, 1969; Maron *et al.*, 1970). In general, there was a good correlation between differences in the amino acid sequence and the immunological activity when the native molecules were compared. However, although no immunological cross-reactivity between native human lysozyme and native hen egg-white lysozyme could be demonstrated, the reduced and alkylated derivatives of these lysozymes cross-react significantly (Arnheim *et al.*, 1971).

Recent studies on the cellular immune response to lysozyme and to CM-lysozyme showed extensive cross-reactivity between these two forms of the molecule, in contrast to the lack of humoral cross-reactivity previously mentioned. The cross-reactivity on the cellular level was demonstrated by delayed skin reactivity, by inhibition of migration of peritoneal exudate cells, and by stimulation of DNA synthesis in splenic lymphocytes (Thompson *et al.*, 1971).

L. Myoglobin

Myoglobin is an oxygen-carrying protein of molecular weight 18,400, of which the amino acid sequence (Edmundson, 1965) and the three-dimensional structure (Kendrew *et al.*, 1961) have been elucidated. The amino acid sequence of sperm whale myoglobin is given in the Appendix. Having no disulfide bridges, a great deal of the structural integrity of the molecule resides in its high content of secondary structure, some 75% of its residues being found in a helical configuration (Kendrew *et al.*, 1961). Much of its antigenic activity can be attributed to its primary and secondary structures. Nevertheless, the tertiary structure of the molecule has been implicated in antigenicity. It has been shown that antibodies to the heme-containing protein could detect differences between the heme protein and the apoprotein (Reichlin *et al.*, 1963). It has also been reported that while antibodies to metmyoglobin (heme protein with ferric iron) could not detect differences between metmyoglobin and apomyoglobin (no heme), the interaction with antibodies to apomyoglobin caused the release of heme from metmyoglobin. It was later demonstrated that antisera to apomyoglobin contained at least two populations of antibodies. One population could not distinguish between apomyoglobin and metmyoglobin, whereas the other population could. Furthermore, it was found that the addition of heme to an immune precipitate consisting of this latter group of antibodies and apomyoglobin resulted in only partial reformation of metmyoglobin, whereas addition of heme to precipitates composed of the former group of antibodies and apomyoglobin resulted in the complete reformation of metmyoglobin (Crumpton, 1966).

Studies on the reactivity of a series of artificial myoglobins, containing various metals in the porphyrin, with antibodies against metmyoglobin revealed that the antigenic activity of an artificial myoglobin with a copper metalloporphyrin was identical to that of the heme protein, whereas the reactivity of all other derivatives was reduced (Atassi, 1967*a*). Additionally, it has been shown that a derivative of apomyoglobin in which the tryptophans at positions 7 and 14 were modified (resulting in profound conformational effects) has drastically reduced antigenic activity (Atassi and Caruso, 1968).

The importance of primary and secondary structures to the antigenicity of the molecule was first indicated by findings that chymotryptic peptides could partially inhibit the reaction between antimyoglobin and antiapomyoglobin and their respective antigens (Crumpton, 1964). The partial or complete isolation of eight peptides from a chymotryptic digest of myoglobin was later reported

(Crumpton and Wilkinson, 1965). Of the eight, seven were shown to possess antigenic activity, although one was very weak. Five of the six strongly active peptides were adjacent to or included corners in the native molecule, while the sixth was the C-terminal heptapeptide of the molecule. The totally inactive peptide was from the N-terminus, which is helical in the intact molecule. It was subsequently shown that the C-terminal heptapeptide (Lys–Glu–Leu–Gly–Tyr–Gln–Gly) and the C-terminal hexapeptide were not as active as a peptide comprising the C-terminal 22 amino acids, which is predominantly helical in the native molecule. However, from investigations with shorter peptides it appeared that while the N-terminal lysine of the C-terminal heptapeptide was not involved in binding, as many as four or five C-terminal amino acids were important (Crumpton, 1967a,b). A study of several synthetic C-terminal peptides indicated that the hepta- and hexapeptides were of equal reactivity with antimyoglobin, whereas the penta- and tetrapeptides were progressively weaker. Analogues which contained phenylalanine or p-methoxyphenylalanine in place of the tyrosine were as active as their native counterpart. It was concluded that the leucine residue was the immunodominant group, as it is situated at a corner in the intact molecule, and, furthermore, that the C-terminal determinant was no smaller, and maybe no larger, than a hexapeptide (Crumpton et al., 1970).

Investigations with tryptic peptides of myoglobin indicated that several peptides, occupying several corners in the three-dimensional configuration of the molecule, as well as the C-terminal hexapeptide, were reactive with antimyoglobin. Removal of the N-terminal heptapeptide had no effect on antigenicity. Cleavage of myoglobin with cyanogen bromide yielded antigenically active fragments comprising residues 1-55, 56-131, and 132-153. The first two precipitated with antimyoglobin and therefore may be assumed to contain two or more determinants. Removal of the N-terminal heptapeptide from fragment 1-55 reduced but did not abolish its precipitating activity. In view of the previously demonstrated nonreactivity of this N-terminal heptapeptide, it is probable that it acts only to stabilize the structure of the rest of the fragment. Of the three fragments, the one comprising residues 56-131 appeared to be the most active (Atassi and Saplin, 1968). Modification of arginines 31 and 45 in fragments 1-55, as well as modification of arginine 139 in fragment 132-153, had no effect on their antigenic activity, whereas modification of arginine 118 in the fragment 56-131 had an appreciable effect. This latter reduction in reactivity correlated well with the reduction in reactivity of the unfragmented molecule when it was modified at arginines 31, 118, and 139, suggesting that of these three arginines, arginine 118 was the only one inolved in the antigenic activity of the intact molecule (Atassi and Thomas, 1969).

Cleavage of myoglobin at the proline residues yielded peptides comprising residues 1-36, 37-87, 1-87, 37-119, and 120-153. Peptide 1-36 was shown to precipitate with antiapomyoglobin and had an activity equal to that of the cyanogen bromide fragment 1-55, suggesting that residues 37-55 were not involved in antigenicity (Atassi and Singhal, 1970). Additionally, peptide 1-31 has been shown to be as reactive with antibodies to myoglobin as was peptide 1-36. Succinylation of lysine 16 of this peptide (1-31) abolished this reactivity, whereas carboxymethylation of the histidine residues at positions 12 and 24 had no effect (Perlstein and Atassi, 1971). Peptide 37-87, while not precipitating with antibodies, has been shown to possess inhibitory activity; with a given antiserum, the sum of the activities of peptides 1-36 and 37-87 was equal to that of fragment 1-87. Fragment 37-119 was also shown to have precipitating activity with antimyoglobin, but peptide 120-153 had only inhibitory activity. The activity of the latter was equal to that of the cyanogen bromide fragment 132-153, suggesting that residues 120-131 were inactive (Atassi and Singhal, 1970).

As has already been indicated, studies with chemically modified derivatives have yielded valuable data concerning the antigenicity of myoglobin. Totally N-acetylated myoglobin could not be shown to react with antimyoglobin, either by complement fixation or by precipitation. It was pointed out that the loss of activity could have been due to conformmational changes and not necessarily due to the loss of amino groups. In addition, treatment of myoglobin with carboxypeptidase removed a C-terminal glutamine and partially removed the penultimate serine, neither of which appeared to be important to antigenic activity (Atassi, 1966). It is probable that this was a maverick myoglobin since the reported C-terminal dipeptide of sperm whale myoglobin is Gln-Gly (rather than Ser-Gln). It is interesting that the C-terminal dipeptide (Ser-Gln) investigated by Atassi did not appear to play a role in the antigenicity, whereas Crumpton (1967a) demonstrated that the dipeptide Gln-Gly did contribute to antigenicity. This indicates that although amino acid changes may result in a change in the general makeup of an area on a protein, this area may still dominate as a determinant group in either form.

Other studies with chemically modified myoglobins have indicated that both methionine residues (Atassi, 1967b, 1969) as well as tryptophan 7 (Atassi and Caruso, 1968) were not involved in the antigenic activity of the protein. Nitration of all three tyrosine residues of myoglobin diminished its antigenic activity. Nitration of the cyanogen bromide fragment 56-131 had no effect on its reactivity with antibodies, but nitration of the fragment 132-153 (at Tyr 146 and Tyr 151) abolished its reactivity (Atassi, 1968).

M. Hemoglobin

Hemoglobin is an oxygen-carrying molecule composed of two pairs of polypeptide chains, each of which carries a heme group. It has an approximate molecular weight of 67,000 (in mammals). The sequence of the a and β chains of human hemoglobin is shown in the Appendix.

Studies on the immunochemical properties of porcine hemoglobin and horse muscle myoglobin demonstrated that rabbit antisera to hemoglobin or myoglobin were able to distinguish between the heme proteins and their globin moieties, indicating the presence of conformation-dependent determinants on these molecules (heme did not directly participate in antigenicity) (Reichlin *et al.*, 1963). It has also been shown that the reaction of antibodies to human hemoglobin with the oxy-form of the molecule was stronger than with its deoxy-form regardless of the form of the immunogen. The difference between the reactivity of the oxy- and the deoxy-forms was most pronounced when anti-horse hemoglobin was assayed with human hemoglobin A_1 (the major normal adult form), suggesting that antibodies were better able to discriminate conformational differences when they were reacted with more distantly related antigen (Reichlin *et al.*, 1964a, 1965a). Hemoglobins containing various other ligands (such as CO) were immunologically indistinguishable from oxyhemo-globin. These observations support chemical evidence which indicates that deoxyhemoglobin has a conformation different from that of liganded hemo-globins, especially with respect to the β chains (Muirhead and Perutz, 1963). The finding that antibodies to native human oxyhemoglobin were able to distinguish between the oxy- and deoxy-forms of two hybrids (containing two canine a chains and two human β chains or vice versa) but could not distinguish one oxy- or deoxy-hybrid from the other oxy- or deoxy-hybrid, respectively, suggested that the a and β chains were equally involved in the conformational changes accompanying oxygenation (Reichlin *et al.*, 1965b).

Rabbit antibodies to human hemoglobin A_1 have been shown to dis-tinguish mutant hemoglobins S and C (differing only in position 6 of the β chains) and hemoglobin H (composed of four β chains) from hemoglobin A_1. The single amino acid difference at position 6 in hemoglobins A_1, S, and C may have altered the binding to antibody via an alteration in the local conformation of the N-terminus, although no profound differences were found between the crystal structures of A_1 and S oxyhemoglobins (Perutz *et al.*, 1951). However, there also exists the possibility that the change at position 6 was accompanied by a conformational change far removed from the N-terminus (Reichlin *et al.*, 1964b). When 18 other mutant human hemoglobins were tested with rabbit antisera to human hemoglobin A_1, the immunological activity of the mutant hemoglobins was inversely related to the extent of differences between the primary sequences of the mutant and rabbit hemoglobins. Six mutants, which differed from A_1 in their reactivity with anti-A_1, differed from A_1 in those regions where human and rabbit hemoglobins contain many differences in primary sequence. Mutants indistinguishable from A_1 by anti-A_1 differed in regions of the polypeptide where the primary sequences of human and rabbit hemoglobins are identical. Thus, it was postulated that the regions containing the differences in primary structure between rabbit and human hemoglobins were the ones to which rabbit antibodies were directed (Reichlin, 1969).

Experiments on the reaction of anti-A_1 hemoglobin with the separated a and β chains showed that intact hemoglobin exhibited some determinants attributed to the a and β chains. It was further shown that either of the chains could express quaternary structure-dependent determinants, regardless of the species from which the complementary chains were derived (Reichlin et al., 1965a). Rabbit antibodies induced by immunization with the isolated a and β chains reacted with the immunogen as well as with intact hemoglobin but did not react with the opposite chain. Furthermore, antibodies to isolated a chains reacted with a chains in either the free form or when complexed with β but were less reactive with a chains which were complexed with γ chains (from fetal hemoglobin), indicating a structural difference between adult hemoglobin $(a_2\beta_2)$ and fetal hemoglobin $(a_2\gamma_2)$ with respect to the a chain. The reaction between antibodies to the β chains and adult hemoglobin was poorer than with isolated β chains, indicating that some determinants of the β chains were lost when the chains complexed with a chains to form $a_2\beta_2$ (Reichlin et al., 1966).

Quantitative studies on the specificity of antibodies to human hemoglobins showed that fetal hemoglobin was much more immunogenic in rabbits than was adult hemoglobin (Reichlin, 1970). This was probably due to the increased immunogenicity of the γ chains, since the actual concentrations of antibodies specific to a chains were equal in antisera to adult and to fetal hemoglobins. Also, in anti-adult hemoglobin there were equal fractions of anti-a-specific and anti-β-specific antibodies, whereas in anti-fetal hemoglobin sera there was a two- to threefold higher concentration of anti-γ-specific antibodies. Since the number of radical differences in primary structure between human and rabbit a and β chains, respectively, is similar (seven differences between the β chains and 11 differences between the a chains), their similar antigenicity is not surprising. However, the human γ chains and rabbit β chain differ to a much greater extent (21 amino acid differences); therefore, the γ chain can be expected to be more immunogenic. However, the amino acids responsible for binding with antibodies may not be the same as those involved in immunogenicity, and, furthermore, only certain regions may be involved in antigenicity.

Whereas native isolated a and β chains have been reported by some workers to be immunogenic (Reichlin et al., 1966, 1970), there are conflicting reports regarding the immunogenicity of denatured a chains, although denatured β chains have always been shown to be immunogenic (Askonas and Smyth, 1964; Ovary, 1964; Kraus and Sassano, 1965).

Studies on the reactivity of isolated tryptic peptides from the a chain with antibodies to denatured a chains indicated that the N-terminal area 1-7, residues

93-99, and the C-terminal area 128-139 were antigenically active. In addition, when a longer tryptic peptide containing residues 1-11 was used, it was found to be more active than area 1-7, suggesting that the longer peptide stabilized a preferred- conformation or possibly increased hydrophobic binding with antibody (Kraus, 1967a,b).

N. Tobacco Mosaic Virus and Its Protein Subunit

The tobacco mosaic virus (TMV) is a rodlike particle of molecular weight 40×10^6 having the dimensions 180 by 3000 Å. The particle is composed of 95% protein and 5% RNA. The protein component (TMVP) consists of 2200 identical subunits. Each subunit is of approximate molecular weight 17,500 and consists of 158 amino acids. The structure of TMV and some immunological aspects related to the virus and to the viral protein have been reviewed (Knight, 1961; Rappaport, 1965).

Although immunological cross-reactions can be demonstrated between the virus and the viral protein, the two possess several distinct antigenic determinants. Thus, following absorption of antivirus with the protein, antibodies remained which still reacted with the virus; absorption of antiprotein with the virus left behind antibodies still capable of reacting with the protein.

While still on the viral rod, the C-terminal threonine residue of the protein subunit can be removed by treating the virus with carboxypeptidase. Although the virus thus treated (CT virus), exhibited no altered infectivity, the treatment apparently altered the antigenic properties of the virus. Anti-TMV absorbed with CT virus retained reactivity with native virus but not with CT virus, whereas anti-CT virus absorbed with native virus retained reactivity with CT virus but not with native virus (Harris and Knight, 1955).

Studies consisting of assays of the capacity of various peptides of the protein (obtained by the action of various enzymes) to inhibit the reaction between the virus and antivirus showed that antibodies to TMV were directed against several areas on the protein. The most efficient inhibitors were peptides comprising sequences 18-23, 62-68, 123-134, 129-131, 142-158, and 153-158, the latter representing the C-terminal portion of the protein (Anderer, 1963a). The fact that peptides representing various regions of the protein were capable of partially inhibiting the precipitin reaction suggested that antibodies to the virus were directed against several areas on the viral protein and implied that these areas were present on the exposed surface of the virus. Moreover, antisera to a synthetic antigen composed of the C-terminal hexapeptide conjugated to a protein carrier precipitated with the virus, although the conjugate failed to precipitate with antivirus (Anderer, 1963b). Further investigations utilizing antisera prepared against the C-terminal penta-, tetra-, and tripeptides (each conjugated to a carrier protein) demonstrated that such antisera were able to inactivate the viral infectivity, the inactivating efficiency of the serum increasing with increases in the length of the peptide used for immunization (Anderer and Schlumberger, 1965).

Although several peptides capable of inhibiting the reaction between TMV and anti-TMV were isolated, there is considerable evidence indicating that the tertiary structure of the protein is of great importance in the antigenicity of TMV (Anderer and Handschuh, 1963; Jeener, 1965; von Sengbusch, 1965; van Regenmortel, 1966). The orderly assembly of the protein subunits on the core RNA (resulting in the viral superstructure) may lead to the expression of new determinants, which induce the production of quaternary structure-dependent antibodies. These determinants may involve areas composed of several inter-acting subunits or arise from conformations of individual subunits induced by the interaction of several subunits. In fact, profound antigenic alterations have been found to occur from the interaction of subunits, even in the absence of viral RNA, which results in aggregation (Rappaport and Zaitlin, 1970).

Studies with tryptic peptides which were reactive with antibodies against the subunit protein (TMVP) indicated a major antigenic area which was apparently not expressed on the virus. This activity was attributed to the tryptic peptide comprising residues 93-112 of the protein (having the sequence Ile-Ile-Glu-Val-Glu-Asn-Gln-Ala-Asn-Pro-Thr-Thr-Ala-Glu-Thr-Leu-Asp-Ala-Thr-Arg) (Benjamini et al., 1964, 1965). Stepwise removal of N-terminal amino acids from this eicosapeptide showed that the binding with anti-TMVP did not appreciably diminish even when five amino acid residues were removed (Young et al., 1966). The synthetic C-terminal decapeptide was later ascertained to constitute the antigenic region of the eicosapeptide (Stewart et al., 1966). Further experiments utilizing synthetic C-terminal peptides (varying in length from dipeptide to the decapeptide) showed that the shortest peptide with demonstrable specific bind-ing with anti-TMVP was the C-terminal pentapeptide (Leu-Asp-Ala-Thr-Arg) (Young et al., 1967). These findings imply that possibly determinants other than the C-terminal pentapeptide might be contained in the decapeptide, or that the pentapeptide represented only a portion of a larger determinant. However, the demonstration that the pentapeptide was able to inhibit, almost completely, the reaction between ^{14}C-labeled N-acetyl decapeptide and anti-TMVP indicated that the pentapeptide was the sole determinant of the decapeptide. Moreover, since the decapeptide was able to inhibit, almost completely, the reaction between ^{14}C-labeled N-acetyl eiciosapeptide and anti-TMVP, it was concluded that the C-terminal pentapeptide was the sole determinant of the eicosapeptide (Benjamini et al., 1968a). Indeed, a synthetic decapeptide, consisting of the native C-terminal pentapeptide with five alanine residues attached N-terminally, exhibited the same degree of binding with anti-TMVP as did the native decapep-tide. These findings indicated a relationship between peptide size and binding and that the enhanced binding with increased peptide length was apparently due to some nonspecific factors. Studies with closely related analogues showed that the C-terminal tripeptide Ala–Thr–Arg was critical for antigenic specificity, whereas the N-terminal dipeptide (Leu–Asp) was responsible for expressing or enhancing the binding of the tripeptide with anti-TMVP. It was further shown that a rough correlation existed between the hydrophobicity of the N-terminal portion of the pentapeptide and binding with anti-TMVP (Young et al., 1968). Whereas the binding of the tripeptide with antibodies could not be demonstrated, ^{14}C-labeled N-octanoyl-Ala–Thr–Arg exhibited strong binding with anti-TMVP. In fact, more octanoyl peptide than native decapeptide was found to bind with antibodies, suggesting that the antigenic specificity of the pentapeptide resided in its C-terminal tripeptide Ala–Thr–Arg, the binding of which was expressed through hydrophobicity N-terminal to this tripeptide (Benjamini et al., 1968b). It has been proposed that the hydrophobic areas enhance the binding through a hydrophobic interaction with antibodies (Karush, 1962; Metzger et al., 1963; Singer, 1965; Benjamini et al., 1969). This implies that in addition to having an area complementary to the antigenic determinant (Ala–Thr–Arg), the antibody combining site may contain hydrophobic areas which interact with the N-terminal hydrophobic area of the peptide. Since the hydrophobic area of the antigen does not have rigid structural restrictions (in the above case, it may consist of leucine, isoleucine, tyrosine, pentaalanine, acetyl-, or octanoyl-), it is conceivable that its interaction with the hydrophobic area on the antibody is essentially nonspecific and serves as an important auxiliary force in the binding between the specific, sterically complementary area of the antigen (Ala–Thr–Arg) and the antibody site. There is, however, a restriction imposed on the hydrophobic interaction, in that the participating hydrophobic groups on the antigen and antibody must be present in the correct position for interaction. This restriction is dictated by the juxtaposition of the antigenic sequence and the complementary antibody area to this sequence.

While anti-TMVP sera produced by all animals tested (which included rabbits, guinea pigs, mice, and sharks) reacted with the decapeptide having the sequence Thr–Thr–Ala–Glu–Thr–Leu–Asp–Ala–Thr–Arg, not all antisera reacted with the C-terminal pentapeptide. Some rabbit antisera reacted with the C-terminal hexapeptide and larger peptides, while sera produced by other rabbits were demonstrated to bind with the C-terminal hepta- or octapeptide (Benjamini et al., 1968a). However, all the antisera tested contained antibodies which bound with N-octanoyl-Ala–Thr–Arg, indicating that TMVP immunization of all of the animals tested led to the production of antibodies with specificity toward the tripeptide Ala–Thr–Arg (Benjamini et al., 1969).

Some antibodies with specificity toward Ala–Thr–Arg possessed sufficient hydrophobic complementarity to effect demonstrable binding with the C-terminal pentapeptide, whereas others were demonstrated to bind only with peptides larger than the C-terminal pentapeptide, perhaps because they did not possess sufficient hydrophobic complementarity for demonstrable interaction with the pentapeptide. However, their specificity to Ala-Thr-Arg was made apparent through increasing the hydrophobicity N-terminal to the peptide. Thus, although a given amount of antibodies produced by an individual rabbit combined with a greater amount of octanoyl-Ala-Thr-Arg than with the decapeptide, equilibrium dialysis experiments revealed that the average association constant of the decapeptide was substantially higher than that of the octanoyl tripeptide but that more antibodies were able to bind with the octanoyl tripeptide (Benjamini, unpublished). From this, it appears that one of the important differences in the response of individuals to immunization with TMVP is expressed in the production of antibodies of different affinities which possess similar antigenic specificities.

The specificity to Ala-Thr-Arg of antibodies produced by different rabbits in response to immunization with the protein TMVP was further investigated by comparing the binding of antibodies with N-octanoyl-Ala-Thr-Arg and closely related octanoylated analogs (Benjamini et al., 1969). The analogs included the N-octanoylated peptides Gly-Thr-Arg, Leu-Thr-Arg, a-aminobutyryl-Thr-Arg, Ala-Gly-Arg, Ala-Ser-Arg, Ala-Leu-Arg, Ala-a-aminobutyryl-Arg, and a-aminobutyryl-a-aminobutyryl-Arg. Of these, only N-octanoyl-a-aminobutyryl-Thr-Arg was found to bind with anti-TMVP produced by some rabbits. These experiments indicated the very high specificity of antibodies to Ala-Thr-Arg and also pointed to some subtle differences in the antibody specificity; of four rabbits tested, antibodies produced by one rabbit bound as much of the octanoyl-a-aminobutyryl-Thr-Arg analogue as octanoyl-Ala-Thr-Arg; those produced by another rabbit bound more of this analogue than the native octanoylated sequence; those produced by another rabbit bound less; and those produced by still another rabbit did not bind with this octanoylated analogue. These findings suggested that in addition to the differences in affinities discussed earlier, there also exist subtle differences in specificities of the antibodies produced by the various rabbits in response to immunization with TMVP.

Attempts to immunize experimental animals with antigenic peptides of TMVP were unsuccessful (Spitler et al., 1970). Consequently, rabbits were immunized with the native peptide and several analogues conjugated to succinylated bovine serum albumin (S-BSA). Analysis of the specificity of antibodies induced by the different conjugates showed that antibodies induced by conjugates of those peptides which had previously been shown to bind with anti-TMVP bound with such peptides, whereas antibodies produced by a conjugate of a peptide which did not bind with anti-TMVP bound only with the homologous peptide. Moreover, TMVP could specifically inhibit the reaction between anti-S-BSA-Leu-Asp-Ala-Thr-Arg and the homologous peptide, whereas it could not inhibit (even partially) the binding between anti-S-BAS-Leu-Asp-Ala-Gly-Arg and the homologous peptide Leu-Asp-Ala-Gly-Arg (Fearney, 1970).

The above results, although admittedly obtained from a small number of analogues, demonstrated that the relationship between the structure of the peptide used for immunization and the specificity of the antibodies produced parallels the relationship between peptide structure and binding with antibodies produced in response to immunization with the native protein. The results suggest that the specificity of the receptor on the immunocompetent cell is very similar (if not identical) to the specificity of the antibody produced.

Investigations of several other parameters of the immune response, utilizing the antigenic peptides of TMVP, their analogues, and several of their derivatives, demonstrated that none of the peptides was immunogenic nor could any of them stimulate, *in vitro*, splenic lymphocytes derived from TMVP-sensitized guinea pigs. However, all of the peptides tested (the native decapeptide representing residues 103-112 of the protein, the pentapeptide having the sequence Leu-Asp-Ala-Thr-Arg, N-(Lys)$_4$ pentapeptide, N-(Lys)$_7$ pentapeptide, N-hexanoyl-Ala-Thr-Arg, and N-butyryl-Ala-Thr-Arg) were able to elicit delayed skin reactions in TMVP-sensitized guinea pigs. Moreover, these peptides, and also N-octanoyl- and N-decanoyl-Ala-Thr-Arg, were able to inhibit the migration of peritoneal exudate cells derived from TMVP-sensitized guinea pigs. The data demonstrated that there was a lack of correlation between the ability to stimulate, *in vitro*, splenic lymphocytes and delayed hypersensitivity but that there was a correlation between the latter and inhibition of migration of peritoneal exudate cells. Moreover, the data indicated that both delayed reactivity and inhibition of migration were independent of carrier specificity (Spitler *et al.*, 1970).

III. DISCUSSION

The contribution of several aspects of molecular structure to immunological specificity has been amply stressed in several reviews (Landsteiner, 1945; Karush, 1962; Kaminski, 1965; Sela, 1966, 1967, 1969; Sela *et al.*, 1967; Kabat, 1966, 1968; Pressman and Grossberg, 1968; Goodman, 1969; Benjamini *et al.*, 1971). The precise interrelationship of physicochemical characteristics in the makeup of antigenic determinants is now beginning to emerge from studies with structurally defined antigens. In the following discussion, we shall attempt to integrate the diverse findings which we have previously summarized into a cohesive picture dealing not only with the specificity of antigen–antibody interactions but also with specificity of interactions at several other levels of the immune response.

A. Reaction with Circulating Antibodies

It has been known for many years that there exist antibodies with specificities dependent upon the quaternary structure of their respective antigens. Such antibodies have been demonstrated in several systems. Thus, antisera to TMV contain antibody populations directed against structures present only when the protein subunits exist in high aggregation states (with or without viral RNA). Similarly, studies with hemoglobin demonstrated that antisera to hemoglobin contain some antibodies with specificity for the α and β chains, the binding of which can only be demonstrated when the two chain types are complexed to form the native hemoglobin molecule. Also, it has been shown that some antibodies against the isolated α chains of hemoglobin can distinguish determinant(s) expressed when the α chain is in the free state or complexed in the $\alpha_2\beta_2$ tetramer but not expressed when the α chain is complexed with γ chain to form $\alpha_2\gamma_2$ (fetal) hemoglobin. It should be stressed that, at the present time, such antibodies can only be referred to as being quaternary structure-dependent, for they could be directed either against an area formed by the juxtaposition of two or more subunits or against a conformation(s) on a single subunit, induced by, and dependent upon, interaction with other subunits.

The participation of tertiary structure in immunological specificity is evident from many observations. It has long been known that little cross-reactivity, if any, exists between native proteins and several of their denatured forms. Thus antibodies to native ribonuclease do not react with performic acid-oxidized ribonuclease. Similarly, antibodies to lysozyme do not react with the reduced and carboxymethylated form of the enzyme. Antibodies are able to distinguish between the oxidized and reduced forms of cytochrome c, confirming chemical evidence of conformational differences between these two forms. Antibodies can distinguish between the oxy- and deoxy-forms of hemoglobin, or myoglobin, which differ in their respective conformations. Other examples of tertiary structure-dependent determinants are those of myoglobins containing metalloporphyrins other than heme and those lost following modification of tryptophan 14 of myoglobin. All of these changes were shown to result in conformational alterations and concomitant reduction in binding with antibodies.

In view of the fact that antisera to globular protein antigens contain antibody populations directed against conformation-dependent determinants, the inhibitory capacity of fragments obtained from such proteins on the reaction between the protein and homologous antibodies is usually limited. This is undoubtedly due to a lack of the proper configuration of the peptide in solution and also may be due to the possibility that conformation-dependent antibodies can recognize determinants consisting of areas composed of different, interacting regions of the polypeptide chain. Nevertheless, the binding of small fragments with antibodies to the whole protein has been demonstrated for many globular proteins. It can be argued that such fragments represent sequence-dependent, rather than conformation-dependent, determinants. On the other hand, the possibility exists that a given sequence will conform to the combining site of an antibody which is actually directed against a conformation containing this sequence. It is also possible for antibodies to be directed against areas consisting of different parts of the primary sequence, brought into proximity by the folding of the polypeptide chain. That antibodies against such areas do exist is indicated by studies with insulin, in which it was shown that some determinants can only be distinguished when the A and B chains are covalently linked to form the native molecule. In addition, antibodies to reduced and carboxymethylated (CM) lysozyme have been shown to bind with the N-terminal and with the C-terminal cyanogen bromide fragments of the molecule. However, each of these fragments is capable of inhibiting the binding of the other with these antibodies, indicating that the immunogen had a preferred conformation with the N- and C-termini in close proximity and that the antibodies are directed against an area composed of both regions of the molecule. These N- and C-terminal peptides are relatively large. Thus their ability to inhibit each other (rather than to bind simultaneously) is probably due to steric hindrance. It is not known whether these antibodies would be able to simultaneously bind short peptides from the N- and C-terminal areas.

Antibody specificity may also be dependent upon the secondary structure of a polypeptide. The major antigenic component of poly-γ-D-glutamic acid is probably a secondary structure of helical nature. Optical rotatory dispersion results, which indicate that the heptapeptide is the lower limit for a preferred conformation in solution, parallel immunological results which show that, compared with small oligomers, the heptapeptide or larger peptides exhibit enhanced binding with antibodies to poly-γ-D-glutamic acid. Another example of the contribution of secondary structure to antigenicity is derived from studies with antibodies to bradykinin. Using various analogues of the peptide, it was demonstrated that whereas amino acid changes which alter side chains or overall charge do not drastically affect binding with antibodies, substitutions resulting in a change in the backbone structure of the peptide adversely affect binding. These examples, in conjunction with the findings which revealed the existence of antigenic determinants at several "corners" of myoglobin, corroborate the existence of secondary structure-dependent determinants shown in other systems (reviewed by Sela, 1969).

In addition to the aforementioned antibodies which react with quaternary, tertiary, and secondary structure-dependent determinants, sequence-dependent antigenic determinants have been demonstrated for many polypeptide and protein antigens. In fact, immunochemical studies of antibodies directed against peptides of relatively low molecular weight (induced with immunogenic conjugates of these peptides) demonstrate immunological activity in structures composed of amino acid sequences which constitute various regions of the peptide. Thus antibodies to ACTH, to gastrin, and to angiotensin have been shown to react with determinants governed by primary amino acid sequence rather than by more complex architectural features of the peptide. Moreover, in these systems the capacity of relatively short, isolated peptides to inhibit the reaction between the native peptide and its antibodies is generally greater than that of isolated peptides from complex globular proteins. However, relatively high antigenic activity has been found to be possessed by some fragments of globular proteins of relatively high molecular weight. Although such reactive peptides (from larger proteins) may be conforming to an antibody site which is actually directed against a conformation containing this peptide, recent studies indicate the presence of sequential determinants in such proteins. Thus immunizations with antigenically active peptides, artificially conjugated to protein carriers, elicit antibodies which react with the native protein.

The discovery of several sequential determinants and their respective antibodies has permitted extensive investigations of the physicochemical characteristics of binding in the absence of the constraints imposed by conformational dependency. It has been demonstrated that the C-terminal tetrapeptide amide of gastrin constitutes the antigenic determinant of the molecule, whereas the contribution of the N-terminal region is only to enhance the binding of this determinant. Moreover, the C-terminal tetrapeptide becomes inactive when the C-terminal amide group is converted to a free carboxyl. An opposite situation is observed with angiotension, where it was demonstrated that the C-terminal carboxyl group is indispensable for antigenicity. Additionally, a phenolic hydroxyl was shown to be important for the antigenic activity of the N-terminal region of this molecule. There are other examples wherein subtle changes in the side chains of amino acids which make up determinants affect, to a greater or lesser extent, their binding with antibodies. A change from valine to asparagine in the antigenically active C-terminal octapeptide of C. pasteurianum ferredoxin greatly reduces reactivity with antibodies; acetylation of lysine 61 in the immunologically active peptide comprising residues 38-61 of ribonuclease greatly reduces its activity with antibodies to oxidized rebonuclease.

The findings with ACTH are interesting from another point of view, namely, that while the C-terminus of the entire molecule (1-39) constitutes a determinant, the removal of residues 25-39 resulted in the expression of antigenicity again in the C-terminus, this time of peptide 1-24. In this latter case, however, it appears that specificity resides N-terminal to the C-terminal tetrapeptide, since peptide 20-24 is not active, whereas peptides 17-24 and 11-24 are active. The greater inhibition of the reaction between peptide 1-24 and antibodies to this peptide which is seen with peptides 17-24 and 11-24 than with peptide 1-24 may be analogous to the observation that anti-DNP antibodies induced by immunization with a-DNP-oligolysines of various chain lengths have two peaks of reactivity when assayed with several antigens. All of the antibodies were shown to be highly reactive with a-DNP-Lys$_2$, but the reactivity decreased with longer polylysine chain lengths and did not increase until chain length of the test polylysyl antigen approached that of the immunogen. Antigens with longer polylysine chains were also less reactive (Schlossman and Levine, 1970).

The above considerations notwithstanding, some changes in the amino acid composition of antigenic peptides do not appreciably affect their binding with antibodies. This may not mean a lack of antigenic specificity but rather implies that the amino acid in question is important only in the complementary fit between the peptide and antibodies and does not directly participate in binding. On the other hand, even though the amino acid may be present in a sequence which binds with antibodies, it may not make any contributions to specificity. Often, the complete removal of the amino acid at this position will only minimally affect the binding.

Some of the above points are illustrated by immunochemical work with an antigenic determinant of tobacco mosaic virus protein. A pentapeptide from this protein (Leu-Asp-Ala-Thr-Arg) binds with antibodies to the whole protein, whereas the C-terminal tetrapeptide is inactive. Studies with analogues demonstrated that substitution of the leucine with glycine, or of the aspartic acid with glutamic acid, abolishes activity, implying that these two N-terminal amino acids are involved in the antigenic specificity of the peptide. However, substitution of these N-terminal amino acids with amino acids of varying degrees of hydrophobicity results in peptides with relative activities correlating with the degree of hydrophobicity. Moreover, substitution of these two N-terminal amino acids with octanoic acid results in an immunologically active peptide with antigenic specificity confined to the C-terminal tripeptide moiety of the octanoylated peptide. Furthermore, except for substitution of the alanine residue in the octanoylated peptide by a-aminobutyric acid, many other substitutions with closely related amino acids yield inactive peptides. The antigenically active, native pentapeptide may thus be viewed as consisting of two functional areas: the C-terminal tripeptide, which is critical for antigenic specificity, and the N-terminal region, which enhances and expresses the binding of the C-terminal tripeptide with antibodies.

It has been suggested that the hydrophobic area enhances the binding through a hydrophobic interaction with antibodies (Karush, 1962; Metzger et al., 1963; Singer, 1965; Benjamini et al., 1969). This implies that, in addition to having an area complementary to the antigenic determinant (Ala-Thr-Arg), the antibody combining site contains hydrophobic area(s) complementary to the hydrophobic area of the antigenic determinant. Since the hydrophobic area of the antigen is not rigidly restricted (with the TMVP it may consist of leucine, isoleucine, tyrosine, pentaalanine, acetyl-, or octanoyl-), it is conceivable that its interaction with the hydrophobic area on the antibody is essentially nonspecific, and serves as an important auxiliary force in the binding between the specific, sterically complementary area of the antigen (Ala-Thr-Arg) and the antibody active site. There is, however, a restriction imposed on the hydrophobic interaction, in that the participating hydrophobic groups on the antigen and antibody must be present in the correct position for interaction. This restriction is dictated by the juxtaposition of the antigenic sequence responsible for specificity and its complementary region on the antibody. It is thus possible to view the binding of the antigenic pentapeptide with antibodies as being governed by an area dictating its antigenic specificity and a hydrophobic area which enhances binding.

It is not meant to imply that this model is universal: with other determinants, both antigenic specificity and binding may be simultaneous properties of one group of residues. Also, auxiliary binding forces may be other than hydrophobic (such as hydrogen bonds and electrostatic interactions) (see Karush, 1962, and Singer, 1965, for reviews). Indeed, when viewed in the perspective of this model, the aforementioned findings of Schlossman and Levine take on new meaning. It is possible, as these workers have proposed, that each a-DNP-oligolysine of different chain length represents a new structure against which a new specificity of antibody is produced. However, it is also possible that the specificity of all antibodies produced in this system is directed at a-DNP-Lys$_2$ but that the binding site in each case is surrounded by acidic areas of varying size and intensity. The model proposed above allows the insertion of any number of specificities into generalized areas of acidity, basicity, or hydrophobicity, of varying size and intensity, thus greatly reducing the amount of genetic information required to generate large numbers of antibodies with different specificities and affinities.

The above discussion has dealt primarily with structure–function relationships in the interaction of antibodies with their respective antigens. An important and interesting corollary of this concerns the differential expression of some determinants. Such differentially expressed determinants, termed "heterotopes" (Bonavida and Sercarz, 1971), always retain their structural integrity and reactivity with antibodies but may or may not induce the formation of antibodies, depending upon the form of the immunogen in which they are located. In addition to the example cited by Bonavida and Sercarz, the isoleucine site of the cytochrome c system may be of this nature. In this case, antibodies produced against monkey cytochrome c (threonine at position 58) do not distinguish between monkey (threonine at position 58) and human (isoleucine at position 58) cytochromes c, whereas antibodies produced against horse cytochrome c (threonine at position 58) are more reactive with monkey than with human cytochrome c, the latter two being identical except at position 58. Thus, the threonine at position 58 is immunosilent in the monkey, but not in the horse, cytochrome c. A somewhat analogous situation may exist at position 47 of cytochrome c. In this case, antibodies against horse cytochrome c (threonine at position 47) do not distinguish between horse and donkey cytochromes c (the latter with serine at position 47), whereas antibodies to human cytochrome c

(serine at position 47) do distinguish between these two. It is possible that this area is immunosilent in the horse protein but not in the human protein.

It appears that one important factor in the specificity of antibodies in a given population is the form of the antigen which reaches the immune machinery. This phenomenon is seen in the generalized higher reactivity of antisera to cytochrome c with the ferri-form of the molecule and the higher reactivity of antisera to hemoglobin with the oxy-form of the molecule, regardless of the form of the immunogen used for immunization. These phenomena suggest that immunogens may be converted *in vivo* to other forms before they are recognized immunologically.

B. Specificity of Cellular Immune Reactions

It is appropriate to begin this part of the discussion with a short summary of some recent, important developments in cellular immunology. It has long been suspected that the thymus plays a central role in the immune response, as indicated by the general immune incompetence of thymectomized animals (reviewed by Miller and Mitchell, 1969; Metzger, 1970; Talmage *et al.*, 1970).

The probable function of the thymus has only recently been brought into perspective by experiments of several workers (Claman *et al.*, 1966; Claman and Chaperon, 1969; Davies, 1969; Miller and Mitchell, 1969). Based on experiments with mice, it is now generally accepted that there are at least two functionally distinct cell types involved in the generation of an immune response leading to antibody synthesis. One of these cell types ("B cells" in current terminology) is derived from the bone marrow and ultimately is directly responsible for the expression of humoral immunity. The other cell type ("T cells" in current terminology) is also derived from the bone marrow but comes under the influence of the thymus at some point prior to its attainment of immune competence. It is generally accepted that the T cell does not secrete immunoglobulins but is nevertheless required for the expression of a humoral response as well as for delayed hypersensitivity. Since both cell types are required for an antibody response, and since both are apparently antigen specific, a cell-cell interaction, bridged by antigen, has been proposed (Mitchison, 1967; Rajewsky *et al.*, 1969). It has been questioned whether this two-cell phenomenon is universal or whether it is operative only in the case of a few antigens. It is beginning to appear, however, that at least the majority of protein antigens do exhibit so called "thymus dependency" (Mitchell, 1971). There is also evidence which indicates that both cell types may be required for the expression of immunological memory (Rajewsky *et al.*, 1969; Katz *et al.*, 1970; Paul *et al.*, 1970). We thus have at our disposal a convenient mechanism whereby one can explain differences in immune specificity at various levels of the immune response.

Based on the two-cell phenomenon, it has recently been postulated that there is a functional division of recognition between the two cell types. With glucagon as a model, it has been proposed that the B cell is specific to a region of the molecule to which predominantly circulating antibodies are eventually directed, while the T cell is specific to a region of glucagon (to which only few antibodies are directed) which may be looked upon as constituting the "carrier" for the antibody-reactive (haptenic) region (Senyk *et al.*, 1971*a,b*).

A functional division within the immune mechanism offers an explanation for the presence of antibodies with specificity to autologous sequences, such as those antibodies described for several of the antigens dealt with in this chapter (ACTH, gastrin, cytochrome *c*, etc.). It is possible that a significant B-cell population exists with specificity toward self-constituents present in the circulation at low concentrations. However, due to the probable presence of either limited numbers or tolerant T cells, the B cells do not become activated. They may, however, be activated by T cells directed against "foreign" carriers to which the autologous sequences are attached, thus leading to the production of autoantibodies. Such a mechanism was proposed for the induction of autoantibodies to thyroglobulin (Chiller *et al.*, 1971). Some of the proteins dealt with herein serve as excellent examples to illustrate this phenomenon. For instance, immunization with heterologous ACTH leads to the production of circulating antibodies, some of which are directed against the area of the molecule (peptide 1-24) which has an identical amino acid sequence in all known ACTHs. Similarly, immunization with gastrin leads to the production of antibodies with specificity toward the *C*-terminal tetrapeptide amide, an area which is common to all known gastrins. In these examples, the areas of the molecule which are not common with those of the immunized animal may serve as "carriers" for the "haptenic" common areas. Existing B cells, recognizing the latter, become activated in the presence of T cells to the foreign regions.

A probable reason for the presence of self-reactive B cells but not such T cells is indicated by recent experiments which showed that T cells are more susceptible to tolerance induction by low antigen concentrations than are B cells (Chiller *et al.*, 1971). However, the presence of self-reactive T cells cannot be ruled out since delayed hypersensitivity to autologous sequences has been demonstrated with peptides of ACTH. It is possible that the cells which mediate delayed hypersensitivity are functionally distinct from those which act as helper cells for the induction of humoral immunity, although both may be thymus derived. This possibility will be discussed later.

Although the specificity of cellular immune reactions has been less well studied than its humoral counterpart, some interesting findings are beginning to emerge. It has been shown that the B chain of insulin is capable of eliciting delayed skin reactions in guinea pigs sensitized to insulin, whereas the A chain is inactive in this respect (Clark and Munoz, 1970). This parallels the humoral reactivity of these chains, where the A chain is seldom reactive with antibodies to insulin. A similar situation is observed with peptide 1-24 of ACTH, where both humoral and delayed reactivity appear to be directed at the C-terminus of the molecule. Additionally, the N-terminal and the C-terminal peptides of oxidized ferredoxin, both reactive with antibodies to oxidized ferredoxin, are both able to elicit delayed skin reactions, and inhibit the migration of peritoneal macrophages, in guinea pigs sensitized to oxidized ferredoxin. This parallelism is also seen in the ability of the peptide which binds with anti-TMVP to elicit delayed reactions in TMVP-sensitive guinea pigs. Another interesting observation is that made with glucagon. In this case, it was shown that not only the N-terminal region of the molecule, which is reactive with antiglucagon, but also the C-terminal region (which is only poorly reactive with these antibodies) can elicit delayed reactions in guinea pigs sensitized to glucagon.

Perhaps the most striking feature of these observations is the apparent absence of "carrier specificity," which is commonly observed in hapten-conjugate systems. It can be argued that the various peptides are large enough to act as their own carriers, but this is difficult to accept in the case of short peptides, such as that of TMVP. Moreover, there have been reports of other small molecules capable of eliciting delayed hypersensitivity reactions (Leskowitz, 1963; Borek, 1968). Several mechanisms have been postulated to account for the apparent lack of delayed reactions to haptens. For example, the binding between a hapten and a receptor may not be sufficient to activate the cell. Also, the receptors may be directed against so-called linkage determinants wherein the carrier contributes to the specificity. However, the apparent lack of carrier specificity in the delayed response to several small peptides may be a moot point, in that "carrier specificity" may be a quantitative rather than qualitative phenomenon. There exists the possibility that the inability to demonstrate delayed reactions to a given hapten is due to an insufficient number of reactive cells specific to the hapten rather than to the absence of such cells. Such a situation may arise due to the availability of a wide variety of "carrier" determinants, in contrast to the single specificity of the hapten. This argument implies that sensitization with a hapten coupled to an autologous carrier (to which no cells are directed) will force the recruitment of hapten-specific cells and result in hapten-specific delayed reactivity. Such hapten-specific delayed reactivity has been observed in the response of humans to DNP-human serum albumin conjugates (Brandriss and Bullock, 1970).

In view of these considerations, the lack of "carrier specificity" in the elicitation of delayed reactions seen in the examples given earlier may be academic: the fact remains that, in a number of cases, delayed reactions can be elicited by small peptide antigens or by peptide fragments derived from protein antigens.

Notwithstanding the above considerations, it is apparent that the ability of isolated peptides to stimulate DNA synthesis in splenic lymphocytes derived from sensitized animals is the exception rather than the rule. This may be due to a requirement for "multifunctionality" in order to stimulate such cells (presumably T cells). This possibility is indicated by the results obtained with ferredoxin, where it has been shown that the simultaneous recognition of at least two antigenic structures is required for stimulation. A possible mechanism whereby "multifunctionality" can be expressed is that in which two receptors on one cell must simultaneously interact with their respective determinants in order to activate the cell. This possibility becomes especially attractive in light of recent experiments which showed that the *in vitro* transformation of lymphocytes by anti-Fab was abolished when these antibodies were used in their monovalent Fab form, indicating that stimulation was dependent upon the simultaneous binding of two cell surface structures by a divalent molecule (Fanger *et al.*, 1970). In view of the results with ferredoxin, and, for that matter, with most immunogenic proteins, which do not exhibit repeated structures, a multifunctional requirement such as this implies that the antigen-sensitive lymphocyte carries receptors of more than one specificity. It is in fact probable that the cell would carry a large variety of specificities, as has been suggested by others (Simonsen, 1967; Miller *et al.*, 1971). The important point is that the overall specificity of the cell would depend on the precise, spatial relationship of each receptor within the milieu of all other receptors. Thus a random distribution of receptors would result in the generation of a large number of precise specificities.

If the above is correct, it may be suggested that the requirements for the stimulation of splenic lymphocytes are different from those for the elicitation of delayed reactions. However, this suggestion is difficult to reconcile with the many findings which indicate that the delayed reaction is more intimately related to the thymus than is the humoral response and thus may be T-cell mediated (Arnason *et al.*, 1962; Aspinall *et al.*, 1963; August *et al.*, 1968; Cooper *et al.*, 1968; DiGeorge, 1968; Good *et al.*, 1968; Meuwissen *et al.*, 1969*a,b*). Nevertheless, the different requirements for elicitation of delayed reactions and for stimulation of lymphocytes are consistent with a model in which some T cells, after the initial contact with immunogen, proliferate and differentiate to become cells, sensitive to single determinants, which mediate delayed reactions. Other T cells involved in the initial, multifunctional contact with immunogen may proliferate, either with or without differentiation, and eventually interact with B cells to generate an antibody response. A mechanism such as this allows delayed hypersensitivity to exist in the absence of humoral immunity, and vice versa. It is tempting to speculate that the role of the macrophage in the immune response is perhaps expressed at the level of the multifunctional stimulation of the T cell. That this may be so is indicated by recent experiments which showed that the presence of macrophages greatly enhances *in vitro* stimulation of lymphocytes (Seeger and Oppenheim, 1970).

In contrast to the above model, it is possible that the delayed reaction is not mediated by thymus-derived cells but rather that it is similar to the humoral response; i.e., the effector substances are elaborated by a B-type cell but require the presence of a T cell at some time prior to the elicitation of the response. It is, however, difficult to accept this idea in view of examples which show absence of humoral reactivity in the presence of delayed reactivity; cross-reactivity with respect to delayed hypersensitivity has been observed between lysozyme and CM-lysozyme, although these two forms do not cross-react on the humoral level (Thompson *et al.*, 1971). A similar phenomenon has been observed with flagellin and acetoacetylated flagellin (Parish, personal communication). These latter phenomena can be explained by proposing that only a few common structures exist between the two forms of the respective molecules and that it is these common structures which are recognized by T cells. Thus, if the cellular interaction which is necessary for a humoral response does involve abridgement by antigen, the B-cell side of this abridgement would rarely, if ever, see the common structures, and cross-reacting antibodies would not be produced even if B cells capable of producing them were present. A one-sided interaction such as this would also explain the frequently observed individual and species differences in the specificity of antibodies produced in response to a given immunogen. In such cases, different animals may utilize different determinants for T-cell recognition. This probably occurs during the response of the rabbit to heterologous cytochromes *c*, where the antibodies produced are predominately directed against areas controlled by sequences common to all of the cytochromes *c*. Presumably, those areas which differ are utilized for recognition by the T cells and are only rarely seen by B cells. It would be informative to investigate the specificity of delayed reactivity in the cytochrome *c* system.

At this point in the discussion, it is appropriate to speculate as to the specificity of the receptor of the antibody-forming cell. Germane to this subject are the results obtained from the characterization of the specificity of serum antibodies elicited in response to immunization with conjugates composed of succinylated bovine serum albumin and several peptides, those which bind with anti-TMVP and those which do not. Although only a limited number of conjugates were tested, the data indicate that the antibodies evoked by conjugates containing peptides which bind with anti-TMVP bind only with peptides which bind with anti-TMVP. Since the B cell is assumed to be associated with the production of hapten-directed antibodies, the above findings may be extrapolated to suggest that the receptor on the B cell possesses specificity similar to that of the antibodies elicited in response to immunization with TMVP. Furthermore, since the specificity of the antibodies elicited in response to immunization with the conjugates paralleled that of the antibodies elicited in response to immunization with TMVP, it may be suggested that the specificities of the antibody produced in response to a given determinant are independent of the carrier on which that determinant is located.

IV. CONCLUDING REMARKS

The immune mechanism is operative at several levels, such as immunogenicity, antigenicity, delayed and immediate hypersensitivity, and tolerogenicity. Several experimental models have been utilized in order to define various parameters of the immune response at one or more of these levels. By focusing our attention on protein antigens of known structure and/or amino acid sequence, we have attempted to interrelate many diverse observations derived from these models. Where some differences in interpretations and conclusions do exist (depending upon the model studied), we have attempted to reconcile the differences and accentuate the common features, in order to emphasize possible underlying principles of immune specificity.

APPENDIX

Poly-γ-D-glutamyl peptide of *Bacillus anthracis:*

$$\underset{\displaystyle (\text{-C-NH-CH-CH}_2\text{-CH}_2\text{-C-NH-CH-CH}_2\text{-CH}_2\text{-C-NH-})_n}{\overset{\displaystyle \underset{\|}{O}\quad \underset{|}{COOH}\quad \underset{\|}{O}\quad \underset{|}{COOH}\quad \underset{\|}{O}}{}}$$

Adrenocorticotropic hormone (ACTH) (human ACTH, according to Lee *et al.,* 1961):

SER- TYR-SER-MET-GLU-HIS-PHE-ARG-TRP-GLY[10]-LYS-PRO-VAL-GLY-LYS-LYS-

ARG- ARG- PRO- VAL- LYS[20]- VAL- TYR- PRO- ASP- ALA- GLY-GLU-ASP-GLN-SER[30]-

ALA-GLU-ALA-PHE-PRO-LEU-GLU-PHE

Angiotensin (horse angiotensin, according to Skeggs *et al.,* 1957):

ASP-ARG-VAL-TYR-ILE-HIS-PRO-PHE-HIS-LEU

Bradykinin (bovine bradykinin, according to Elliot *et al.,* 1960):

ARG-PRO-PRO-GLY-PHE-SER-PRO-PHE-ARG

Gastrin (human gastrin, according to Bentley *et al.,* 1966):

GLU- GLY- PRO- TRP- LEU- GLU- GLU- GLU- GLU-GLU[10]-ALA-TYR-GLY-TRP-MET-

ASP-PHE

Glucagon (bovine glucagon, according to Bromer *et al.*, 1957):

10

HIS- SER- GLN- GLY- THR-PHE-THR-SER-ASP-TYR-SER-LYS-TYR-LEU-ASP-SER-

20

ARG-ARG-ALA-GLN-ASP-PHE-VAL-GLN-TRP-LEU-MET-ASN-THR

Ferredoxin (*Clostridium pasteurianum* ferredoxin, according to Tanaka *et al.*, 1966):

10

ALA- TYR- LYS- ILE-ALA-ASP-SER-CYS-VAL-SER-CYS-GLY-ALA-CYS-ALA-SER-

20   30

GLU- CYS- PRO- VAL- ASN- ALA- ILE-SER-GLN-GLY-ASP-SER-ILE-PHE-VAL-ILE-

40

ASP- ALA- ASP-THR- CYS-ILE-ASP-CYS-GLY-ASN-CYS-ALA-ASN-VAL-CYS-PRO-

50

VAL-GLY-ALA-PRO-VAL-GLN-GLU

Insulin (human insulin, according to Nicole and Smith, 1960):

A chain

10

GLY-ILE-VAL-CLU-GLN-CYS-CYS-THR-SER-ILE-CYS-SER-LEU-TYR-GLN-LEU-

20

GLU-ASN-TYR-CYS-ASN

B chain

10

PHE- VAL- ASN-GLN-HIS-LEU-CYS-GLY-SER-HIS-LEU-VAL-GLU-ALA-LEU-TYR-

20

LEU-VAL-CYS-GLY-GLU-ARG-GLY-PHE-PHE-TYR-THR-PRO-LYS-THR

Cytochrome *c* (Human cytochrome *c*, according to Matsubara and Smith, 1962):

10

GLY- ASP- VAL- GLU- LYS- GLY-LYS-LYS-ILE-PHE-ILE-MET-LYS-CYS-SER-GLN-

20   30

CYS- HIS- THR- VAL-GLU-LYS-GLY-GLY-LYS-HIS-LYS-THR-GLY-PRO-ASN-LEU-

40

HIS- GLY- LEU- PHE-GLY- ARG- LYS- THR-GLY-GLN- ALA- PRO-GLY-TYR-SER-

50   60

TYR-THR- ALA- ALA- ASN- LYS-ASN-LYS-GLY-ILE-ILE-TRP-GLY-GLU-ASP-THR-

70

LEU-MET-GLU-TYR-LEU-GLU-ASN-PRO-LYS-LYS-TYR-ILE-PRO-GLY-THR-LYS-

80   90

MET- ILE-PHE- VAL-GLY- ILE- LYS- LYS- LYS- GLU-GLU-ARG-ALA-ASP-LEU-ILE-

100

ALA-TYR-LEU-LYS-LYS-ALA-THR-ASN-GLU

Ribonuclease (Bovine RNase, according to Smyth *et al.*, 1963):

10

LYS- GLU-THR-ALA-ALA-ALA-LYS-PHE-GLU-ARG-GLN-HIS-MET-ASP-SER-SER-

20   30

THR-SER-ALA-ALA-SER-SER-SER-ASN-TYR-CYS-ASN-GLN-MET-MET-LYS-SER-

40

ARG- ASN- LEU-THR-LYS-ASP-ARG-CYS-LYS-PRO-VAL-ASN-THR-PHE-VAL-HIS-

50   60

GLU- SER- LEU- ALA- ASP- VAL- GLN- ALA- VAL- CYS- SER-GLN- LYS-ASN-VAL-

70

ALA- CYS- LYS- ASN- GLY- GLN- THR- ASN- CYS-TYR-GLN-SER-TYR-SER-THR-

80   90

MET- SER- ILE- THR-ASP-CYS-ARG-GLU-THR-GLY-SER-SER-LYS-TYR-PRO-ASN-

100   110

CYS- ALA-TYR- LYS- THR- THR-GLN- ALA- ASN-LYS-HIS-ILE-ILE-VAL-ALA-CYS-

120

GLU-GLY-ASN-PRO-TYR-VAL-PRO-VAL-HIS-PHE-ASP-ALA-SER-VAL

Lysozyme (Hen egg-white lysozyme, according to Canfield, 1963):

                                          10

LYS- VAL- PHE- GLY- ARG- CYS- GLU- LEU- ALA- ALA- ALA- MET- LYS- ARG- HIS-

               20   30

GLY- LEU- ASP- ASN- TYR- ARG- GLY- TYR- SER- LEU-GLY-ASN-TRP-VAL-CYS-

                                      40

ALA- ALA- LYS- PHE- GLU- SER- ASN-PHE-ASN-THR-GLN-ALA-THR-ASN-ARG-

                50   60

ASN- THR- ASP- GLY- SER- THR-ASP-TYR-GLY-ILE-LEU-GLN-ILE-ASN-SER-ARG-

                                      70

TRP-TRP-CYS-ASN-ASP-GLY-ARG-THR-PRO-GLY-SER-ARG-ASN-LEU-CYS-ASN-

             80   90

ILE- PRO- CYS- SER- ALA- LEU- LEU- SER- SER- ASP- ILE- THR-ALA-SER-VAL-ASN-

                          100

CYS- ALA- LYS- LYS- ILE- VAL- SER-ASP-GLY-ASP-GLY-MET-ASN-ALA-TRP-VAL-

110   120

ALA- TRP- ARG- ASN- ARG- CYS- LYS- GLY- THR- ASP- VAL- GLN-ALA-TRP-ILE-

ARG-GLY-CYS-ARG-LEU

Myoglobin (Sperm whale myoglobin, according to Edmundson, 1965):

                                          10

VAL-LEU-SER-GLU-GLY-GLU-TRP-GLN-LEU-VAL-LEU-HIS-VAL-TRP-ALA-LYS-

               20   30

VAL-GLU- ALA- ASP- VAL- ALA-GLY-HIS-GLY-GLN-ASP-ILE-LEU-ILE-ARG-LEU-

                                      40

PHE- LYS- SER- HIS- PRO-GLU-THR- LEU-GLU- LYS-PHE- ASP- ARG-PHE-LYS-HIS-

                50   60

LEU- LYS-THR-GLU- ALA-GLU-MET-LYS-ALA-SER-GLU-ASP-LEU-LYS-LYS-HIS-

                                      70

GLY- VAL- THR- VAL- LEU- THR- ALA- LEU- GLY- ALA- ILE-LEU-LYS-LYS-LYS-

             80   90

GLY- HIS- HIS- GLU- ALA-GLU- LEU-LYS-PRO-LEU-ALA-GLN-SER-HIS-ALA-THR-

                  100   110

LYS-HIS-LYS-ILE-PRO-ILE-LYS-TYR-LEU-GLU-PHE-ILE-SER-GLU-ALA-ILE-ILE-

                                 120

HIS- VAL- LEU- HIS- SER- ARG-HIS-PRO-GLY-ASN-PHE-GLY-ALA-ASP-ALA-GLN-

            130   140

GLY-ALA-MET-ASN-LYS-ALA-LEU-GLU-LEU-PHE-ARG-LYS-ASP-ILE-ALA-ALA-

                          150

LYS-TYR-LYS-GLU-LEU-GLY-TYR-GLN-GLY

Hemoglobin (human hemoglobin, according to Braunitzer *et al.*, 1961):

                                       α chain

                                          10

VAL-LEU-SER-PRO-ALA-ASP-LYS-THR-ASN-VAL-LYS-ALA-ALA-TRP-GLY-LYS-

               20   30

VAL-GLY- ALA- HIS- ALA -GLY- GLU- TYR-GLY-ALA-GLU-ALA-LEU-GLU-ARG-

                                      40

MET- PHE- LEU- SER- PHE-PRO-THR-THR-LYS-THR-TYR-PHE-PRO-HIS-PHE-ASP-

                50

LEU- SER- HIS-GLY-SER-ALA-GLN-VAL-LYS-GLY-HIS-GLY-LYS-LYS-VAL-ALA-

                                      70

ASP- ALA-LEU-THR-ASN-ALA-VAL-ALA-HIS-VAL-ASP-ASP-MET-PRO-ASN-ALA-

            80   90

LEU- SER- ALA- LEU- SER- ASP-LEU-HIS-ALA-HIS-LYS-LEU-ARG-VAL-ASP-PRO-

                  100   110

VAL- ASN-PHE-LYS-LEU-LEU-SER-HIS-CYS-LEU-LEU-VAL-THR-LEU-ALA-ALA-

                                 120

HIS- LEU- PRO- ALA-GLU-PHE-THR-PRO-ALA-VAL-HIS-ALA-SER-LEU-ASP-LYS-

            130   140

PHE-LEU-ALA-SER-VAL-SER-THR-VAL-LEU-THR-SER-LYS-TYR-ARG

β chain

10
VAL-HIS-LEU-THR-PRO-GLU-GLU-LYS-SER-ALA-VAL-THR-ALA-LEU-TRP-GLY
20   30
- LYS- VAL- ASN- VAL- ASP- GLU- VAL-GLY-GLY-GLU-ALA-LEU-GLY-ARG-LEU-
40
LEU-VAL-VAL-TYR-PRO-TRP-THR-GLN-ARG-PHE-PHE-GLU-SER-PHE-GLY-ASP
50   60
- LEU-SER-THR-PRO-ASP-ALA-VAL-MET-GLY-ASN-PRO-LYS-VAL-LYS-ALA-HIS
70
-GLY- LYS-LYS-VAL-LEU-GLY-ALA-PHE-SER-ASP-GLY-LEU-ALA-HIS-LEU-ASP-
80   90
ASN- LEU- LYS-GLY-THR-PHE-ALA-THR-LEU-SER-GLU-LEU-HIS-CYS-ASN-LYS-
100   110
LEU-HIS-VAL-ASP-PRO-GLU-ASN-PHE-ARG-LEU-LEU-GLY-ASN-VAL-LEU-VAL-
120
CYS- VAL- LEU-ALA-HIS-HIS-PHE-GLY-LYS-GLU-PHE-THR-PRO-PRO-VAL-GLN-
130   140
ALA- ALA- TYR- GLN- LYS- VAL- VAL- ALA-GLY-VAL-ALA-ASN-ALA-LEU-ALA-

HIS-LYS-TYR-HIS

Tobacco mosaic virus protein (TMVP) (strain vulgare, according to Anderer *et al.*, 1960; Tsugita *et al.*, 1960):

10
SER- TYR- SER- ILE- THR- THR-PRO-SER-GLN-PHE-VAL-PHE-LEU-SER-SER-ALA-
20   30
TRP- ALA- ASP- PRO- ILE-GLU- LEU- ILE-ASN-LEU-CYS-THR-ASN-ALA-LEU-GLY-
40
ASN-GLN-PHE-GLN- THR-GLN-GLN- ALA- ARG- THR- VAL- VAL-GLN-ARG-GLN-
50   60
PHE- SER- GLN- VAL- TRP- LYS- PRO- SER- PRO- GLN- VAL-THR-VAL-ARG-PHE-
70
PRO- ASP- SER-ASP-PHE-LYS-VAL-TYR-ARG-TYR-ASN-ALA-VAL-LEU-ASP-PRO-
80   90
LEU- VAL-THR-ALA-LEU-LEU-GLY-ALA-PHE-ASP-THR-ARG-ASN-ARG-ILE-ILE-
100
GLU- VAL- GLU- ASN-GLN-ALA- ASN-PRO-THR- THR- ALA-GLU-THR- LEU- ASP-
110   120
ALA- THR- ARG- ARG- VAL- ASP- ASP- ALA-THR- VAL- ALA- ILE- ARG- SER- ALA-
130   140
ILE- ASN- ASN- LEU- ILE- VAL-GLU-LEU-ILE-ARG-GLY-THR-GLY-SER-TYR-ASN-
150
ARG- SER- SER- PHE-GLU-SER-SER-SER-GLY-LEU-VAL-TRP-THR-SER-GLY-PRO-

ALA-THR

REFERENCES

Anastasi, A., Erspamer, V., and Endean R. (1968). *Arch. Biochem.* 125:57.
Anderer, F. A. (1963a). *Z. Naturforsch.* 18b:1010.
Anderer, F. A. (1963b). *Biochim. Biophys. Acta* 71:246.
Anderer, F. A., and Handschuh, D. (1963). *Z. Naturforsch.* 18b:1015.
Anderer, F. A., and Schlumberger, H. D. (1965). *Biochim. Biophys. Acta* 97:503.
Anderer, F. A., Uhlig, H., Weber, E., and Schramm, G. (1960). *Nature* 186:922.
Anderson, J. C., Barton, R. A., Gregory, R. A., Hardy, P. H., Kenner, G. W., MacLeod, J. K.,
    Preston, J., Sheppard, R. C., and Morley, J. S. (1964). *Nature* 204:933.
Anfinsen, C. B. (1962). *Brookhaven Symp. Biol.* 14:184.

Arnason, B. G., Jankovic, D. B., Waksman, B. H., and Wennersten, C. (1962). *J. Exptl. Med.* **116**:177.
Arnheim, N., and Wilson, A. C. (1967). *J. Biol. Chem.* **242**:3951.
Arnheim, N., Prager, E. M., and Wilson, A. C. (1969). *J. Biol. Chem.* **244**:2085.
Arnheim, N., Sobel, J., and Canfield, R. (1971). In preparation.
Arnon, R. (1968). *Europ. J. Biochem.* **5**:583.
Arnon, R., and Sela, M. (1969). *Proc. Natl. Acad. Sci.* **62**:163.
Arquilla, E. R., (1962). *Endocrinology* **14**:146.
Arquilla, E. R., Ooms, H., and Finn, J. (1966). *Diabetologia* **2**:1.
Arquilla, E. R., Ooms, H., and Mercola, D. (1968). *J. Clin. Invest.* **47**:474.
Arquilla, E. R., Bromer, W. W., and Mercola, D. (1969). *Diabetes* **18**:193.
Askonas, B. A., and Smyth, D. G. (1964). *Nature* **201**:496.
Aspinall, R. L., Meyer, R. K., Graetzer, M. A., and Wolfe, H. R. (1963). *J. Immunol.* **90**:872.
Atassi, M. S. (1966). *Nature* **209**:1209.
Atassi, M. Z. (1967*a*). *Biochem. J.* **103**:29.
Atassi, M. Z. (1967*b*). *Biochem. J.* **102**:478.
Atassi, M. Z. (1968). *Biochemistry* **7**:3078.
Atassi, M. Z. (1969). *Immunochemistry* **6**:801.
Atassi, M. Z., and Caruso, D. R. (1968). *Biochemistry* **7**:699.
Atassi, M. Z., and Saplin, B. J. (1968). *Biochemistry* **7**:688.
Atassi, M. Z., and Singhal, R. P. (1970). *Biochemistry* **9**:3854.
Atassi, M. Z., and Thomas, A. V. (1969). *Biochemistry* **8**:3385.
August, C. S., Rosen, F. S., Filler, R. M., Janeway, C. A., Markowski, B., and Ray, H. E. M. (1968). *Lancet* **2**:1210.
Avey, H. P., Carlisle, C. H., and Shukla, P. D. (1962). *Brookhaven Symp. Biol.* **14**:199.
Axelrod, A. E., Traketellis, A. C., and Hofmann, K. (1963). *Nature* **197**:146.
Benjamini, E., Young, J. D., Shimizu, M., and Leung, C. Y. (1964). *Biochemistry* **3**:1115.
Benjamini, E., Young, J. D., Peterson, W. J., Leung, C. Y., and Shimizu, M. (1965). *Biochemistry* **4**:2081.
Benjamini, E., Shimizu, M., Young, J. D., and Leung, C. Y. (1968*a*). *Biochemistry* **7**:1253.
Benjamini, E., Shimizu, M., Young, J.D., and Leung, C. Y. (1968*b*). *Biochemistry* **7**:1261.
Benjamini, E., Shimizu, M., Young, J. D., and Leung, C. Y. (1969). *Biochemistry* **8**:2242.
Benjamini, E., Michaeli, D., and Young, J. D. (1971). In Sela, M. (ed.), *Current Topics in Microbiology and Immunology,* Springer-Verlag, Heidelberg (in press).
Bentley, P. H., Kenner, G. W., and Sheppard, R. C. (1966). *Nature* **209**:583.
Berson, S. A., and Yalow, R. S. (1959). *J. Clin. Invest.* **38**:2019.
Berson, S. A., and Yalow, R. S. (1963). *Science* **139**:844.
Blake, C. C. F., Koenig, D. F., Mair, G. A., North, A. C. T., Phillips, D. C., and Sarma, V. R. (1965). *Nature* **206**:757.
Bonavida, B. (1968). Ph.D. thesis, University of California, Los Angeles.
Bonavida, B., and Sercarz, E. E. (1971). *Europ. J. Immunol.* **I**:166.
Bonavida, B., Miller, A., and Sercarz E. E. (1969). *Biochemistry* **8**:968.
Borek, F. (1968). In Sela, M. (ed.), *Current Topics in Microbiology and Immunology,* Vol. 43, Springer-Verlag, Heidelberg, p. 126.
Boyd, B. W., and Peart, W. S. (1968). *Lancet* **II**:129.
Brandriss, M. W., and Bullock, W. E. (1970). *J. Immunol.* **105**:1416.
Branster, M., and Cinader, B. (1961). *J. Immunol.* **87**:18.
Braunitzer, G., Gehring-Müller, R., Hilschmann, N., Hilse, K., Hobom, G., Rudloff, V., and Wittman-Liebold, B. (1961). *Z. Physiol. Chem.* **325**:283.
Bromer, W. W., Staub, A., Sinn, L. G., and Behrens, O. K. (1957). *J. Am. Chem. Soc.* **79**:2807.
Brown, R. K. (1962). *J. Biol. Chem.* **237**:1162.

Brown, R. K. (1963). *Ann. N.Y. Acad. Sci.* 103:754.
Brown, R. K., Delaney, R., Levine, L., and Van Vunakis, H. (1959a). *J. Biol. Chem.* 234:2043.
Brown, R. K., Durieux, J., Delaney, R., Leikhim, E., and Clark, B. J. (1959b). *Ann. N.Y. Acad. Sci.* 81:524.
Brown, R. K., Tacey, B. C., and Anfinsen, C. B. (1960). *Biochim. Biophys. Acta* 39:528.
Brown, R. K., McEwan, M., Mikoryak, C. A., and Polkowski, J. (1967). *J. Biol. Chem.* 242:3007.
Brunfeldt, K., Hansen, B. A., and Jorgensen, K. R. (1968). *Acta Endocrinol.* 57:307.
Canfield, R. E. (1963). *J. Biol. Chem.* 238:2698.
Canfield, R. E., and Liu, A. K. (1968). *J. Biol. Chem.* 240:1997.
Catt, K., and Coghlan, J. P. (1967). *J. Exptl. Biol. Med. Sci.* 45:269.
Chiller, J. M., Habicht, G. S., and Weigle, W. O. (1971). *Science* 171:813.
Christensen, J. A., Gerwing Levy, J., and Kelly, B. (1971). *Biochemistry* (in press).
Cinader, B. (1967). In *Antibodies to Biologically Active Molecules*, Pergamon Press, Oxford, p. 85.
Cinader, B., and Lafferty, K. J. (1963). *Ann. N.Y. Acad. Sci.* 103:653.
Claman, H. N., and Chaperon, E. A. (1969). *Transplant. Rev.* 1:92.
Claman, H. N., Chaperon, E. A., and Triplett, R. F. (1966). *J. Immunol.* 97:828.
Clark, C., and Munoz, J. (1970). *J. Immunol.* 105:574.
Cooper, M. D., Percy, D. Y., Peterson, R. D. O., Gabrielson, A. E., and Good, R. A. (1968). In *Birth Defects, Immunologic Deficiency Diseases in Man*, National Foundation for the March of Dimes, p. 7.
Crestfield, A. M., Stein, W. H., and Moore, S. (1962). *Arch. Biochem. Biophys. Suppl.* 1:217.
Crumpton, M. J. (1964). *Biochem. J.* 91:40.
Crumpton, M. J. (1966). *Biochem. J.* 100:223.
Crumpton, M. J. (1967a). *Nature* 245:17.
Crumpton, M. J. (1967b). In Cinader, B. (ed.), *Antibodies to Biologically Active Molecules*, Pergamon Press, Oxford, pp. 61.
Crumpton, M. J., and Wilkinson, J. M. (1965). *Biochem. J.* 94:545.
Crumpton, M. J., Law, H. D., and Strong, R. C. (1970). *Biochem. J.* 116:923.
d'Auriac, G. A., Meyer, P., and Milliez, P. (1969). *Compt. Rend. Acad. Sci. Paris Ser. D* 268:1886.
Davidson, J. K., and Haist, R. E. (1965). *Can. J. Physiol. Pharmacol.* 43:373.
Davidson, J. K., Zeigler, M., and Haist, R. E. (1968). *Diabetes* 17:8.
Davidson, J. K., Zeigler, M., and Haist, R. E. (1969). *Diabetes* 18:212.
Davies, A. J. S. (1969). *Transplant. Rev.* 1:43.
Dayhoff, M. O., and Eck, R. V. (1969). *Atlas of Protein Sequence and Structure.*
Deodhar, S. D. (1960). *J. Exptl. Med.* 111:429.
De Zoeten, L. W., and De Bruin, O. A. (1961). *Rec. Trav. Chim.* 80:907.
De Zoeten, L. W., and Havinga, E. (1961). *Rec. Trav. Chim.* 80:917.
Dietrich, F. M. (1967). *Immunochemistry* 4:65.
Di George, A. M. (1968). In *Birth Defects, Immunologic Deficiency Diseases in Man*, National Foundation for the March of Dimes, p. 116.
Edmundson, A. B. (1965). *Nature* 205:883.
Elliot, D. F., Lewis, G. P., and Horton, E. W. (1960). *Biochem. Biophys. Res. Commun.* 3:87.
Fanger, M. W., Hart, D. A., Wells, S. V., and Nisonoff, A. (1970). *J. Immunol.* 105:1484.
Fearney, F. J. (1970). M.S. thesis, San Francisco State College, San Francisco, California.
Felber, J. P., and Micheli, A. (1967). In Cinader, B. (ed.), *Antibodies to Biologically Active Molecules*, Pergamon Press, Oxford, pp. 277.
Felber, J. P., Ashcroft, S. H. J., Villanueva, A., and Vannotti, A. (1966). *Nature* 211:654.
Finn, F. M., and Hofmann, K. (1965). *J. Am. Chem. Soc.* 87:645.

Fischer, K., Hachmeister, V., and Kracht, J. (1965). *Naturwissenschaften* 52:347.
Fleischer, N., Givens, J. R., Abe, K., and Nicholson, W. E. (1965). *J. Clin. Invest.* 44:1047.
Fleischer, N., Givens, J. R., Abe, K., Nicholson, W. E., and Liddle, G. W. (1966). *Endocrinology* 78:1067.
Fujio, H., Kishiguchi, S., Shinka, S., Saiki, Y., and Amano, T. (1959). *Biken J.* 2:55.
Fujio, H., Saiki, Y., Imanishi, M., Shinka, S., and Amano, T. (1962). *Biken J.* 5:201.
Fujio, H., Imanishi, M., Nishioka, K., and Amano, T. (1968*a*). *Biken J.* 11:207.
Fujio, H., Imanishi, M., Nishioka, K., and Amano, T. (1968*b*). *Biken J.* 11:219.
Gelzer, J. (1968). *Immunochemistry* 5:23.
Gerwing, J., and Thompson, K. (1968). *Biochemistry* 7:3888.
Good, R. A., Peterson, R. D. A., Perey, D. Y., Finstad, J., and Cooper, M. D. (1968). In *Birth Defects, Immunologic Deficiency Diseases in Man*, National Foundation for the March of Dimes, p. 17.
Goodfriend, T., Fasman, G., Kemp, D., and Levine, L. (1966). *Immunochemistry* 3:223.
Goodfriend, T. L., Levine, L., and Fasman, G. D. (1964). *Science* 144:1344.
Goodman, J. W. (1969). *Immunochemistry* 6:139.
Goodman, J. W., Nitecki, D. E., and Stoltenberg, I. M. (1968). *Biochemistry* 7:706.
Gregory, R. A. (1966). *Gastroenterology* 51:953.
Grodsky, G. M. (1965). *Rep. Ross. Pediat. Res. Conf.* 51:8.
Grodsky, G. M., Peng, C. T., and Forsham, P. H. (1959). *Arch. Biochem. Biophys.* 81:264.
Gross, E., and Witkop, B. (1962). *J. Biol. Chem.* 237:1856.
Haber, E., Page, L. B., and Jacoby, G. A. (1965). *Biochemistry* 4:693.
Haber, E., Richards, F. F., Spragg, J., Austen, K. F., Valloton, M., and Page, L. B. (1967). *Cold Spring Harbor Sym. Quant. Biol.* 32:299.
Halikis, D. N., and Arquilla, E. R. (1961). *Diabetes* 10:142.
Harris, J. I., and Knight, C. A. (1955). *J. Biol. Chem.* 214:215.
Hedwall, P. R. (1968). *Brit. J. Pharmacol.* 34:623.
Hollemans, J., Van Der Meer, J., and Touber, J. L. (1968). *Nature* 217:277.
Imura, H., Sparks, L. L., Gordsky, G. M., and Forsham, P. H. (1965). *J. Clin. Endrocrinol. Metab.* 25:1361.
Jeener, R. (1965). *Virology* 26:10.
Kabat, E. A. (1968). *Structural Concepts in Immunology and Immunochemistry*, Holt, Rinehart, and Winston, Inc., New York.
Kabat, E. A. (1966). *J. Immunol.* 97:1.
Kaminski, M. (1965). In *Progress in Allergy*, Vol. 9, S. Karger, Basel, pp. 79.
Kartha, G., Bello, J., and Harker, D. (1967). *Nature* 213:862.
Karush, F. (1962). *Adv. Immunol.* 2:1.
Katz, D. H., Paul, W. E., Goidl, E. A., and Benacerraf, B. (1970). *J. Exptl. Med.* 132:261.
Kelly, B., and Gerwing Levy, J. (1971). *Biochemistry* 10:1763.
Kendrew, J. C., Watson, H. C., Strandberg, B. E., Dickerson, R. E., Phillips, D. C., and Shore, V. C. (1961). *Nature* 190:666
Kerp, L., Kasemir, H., Kieling, F., and Steinhilber, S. (1968). *Verh. Gent. Ges. Inn. Med.* 74:552.
Kerp, L., Kasemir, H., Kieling, F., and Steinhilber, S. (1969). *Internat. Arch. Allergy* 36:143.
Kitagawa, M., Onoue, K., Okamura, Y., Anai, M., and Yamamura, Y. (1960). *J. Biochem.* 48:483.
Knight, C. A. (1961). In Heidelberger, M., Plescia, O. J., and Day, R. A. (eds.), *Immunochemical Approaches to Problems in Microbiology*, Rutgers University Press, New Brunswick, N.J., p. 161.
Kraus, L. M. (1967*a*). *Automation in Analytical Chemistry*, Technicon Symposium 1966, pp. 424.
Kraus, L. M. (1967*b*). *J. Immunol.* 99:894.

Kraus, L. M., and Sassano, F. G. (1965). *Proc. Soc. Exptl. Biol. Med.* 118:1037.
Landsteiner, K. (1945). *The Specificity of Serological Reactions,* 2nd ed., Harvard University Press, Cambridge, Mass.
Lee, T. H., Lerner, A. B., and Buettner-Janusch, V. (1961). *J. Biol. Chem.* 236:2970.
Leskowitz, S. (1963). *J. Exptl. Med.* 117:909.
Lockwood, D. H., and Prout, T. E. (1962). *Clin. Res.* 10:401.
Margoliash, E., and Schejter, A. (1966). *Adv. Protein Chem.* 21:113.
Margoliash, E., Reichlin, M., and Nisonoff, A. (1967*a*). In Ramachandran, G. N. (ed.), *Conformation of Biopolymers,* Academic Press, New York, p. 253.
Margoliash, E., Reichlin, M., and Nisonoff, A. (1967*b*). In *Structure and Function of Cytochromes,* pp. 269.
Margoliash, E., Nisonoff, A., and Reichlin, M. (1970). *J. Biol. Chem.* 245:931.
Maron, E., Arnon, R., Sela, M., Perin, J. P., and Jolles, P. (1970). *Biochim. Biophys. Acta* 214:222.
Maron, I., Shiozawa, C., Arnon, R., and Sela, M. (1971). *Biochemistry* 10:763.
Matsubara, H., and Smith, E. L. (1962). *J. Biol. Chem.* 237:3575.
May, J. E., and Brown, R. K. (1968). *Immunochemistry* 5:79.
McGuigan, J. E. (1967). *Gastroenterology* 53:697.
McGuigan, J. E. (1968*a*). *Gastroenterology* 54:1005.
McGuigan, J. E. (1968*b*). *Gastroenterology* 54:1012.
McGuigan, J. E. (1969*a*). *Gastroenterology* 56:429.
McGuigan, J. E. (1969*b*). *Gastroenterology* 56:858.
Merigan, T. C., and Potts, J. T. (1966). *Biochemistry* 5:910.
Metzger, H. (1970). *Ann. Rev. Biochem.* 39:889.
Metzger, H., Wofsy, L., and Singer, S. J. (1963). *Arch. Biochem. Biophys.* 103:206.
Meuwissen, H. J., Van Alten, P. A., and Good, R. A. (1969*a*). *J. Immunol.* 102:1079.
Meuwissen, H. J., Van Alten, P. A., and Good, R. A. (1969*b*). *Transplantation* 7:1.
Miller, A., Deluca, D., Decker, J., Ezzell, R., and Sercarz, E. (1971). *Am. J. Pathol.* 65:451.
Miller, J. F. A. P., and Mitchell, G. F. (1969). *Transplant. Rev.* 1:3.
Mills, J. A., and Haber, E. (1963). *J. Immunol.* 91:536.
Mitchell, B., Gerwing Levy, J., and Nitz, R. (1970). *Biochemistry* 9:1837.
Mitchison, N. A. (1967). *Cold Spring Harbor Symp. Quant. Biol.* 32:431.
Moloney, P. J., and Coval, M. (1955). *Biochem. J.* 59:179.
Muirhead, H., and Perutz, M. F. (1963). *Nature* 199:633.
Mutt, V., and Jorpes, J. E. (1967). *Biochem. Biophys. Res. Commun.* 26:392.
Nicol, D. S. H. W., and Smith, L. F. (1960). *Nature* 187:483.
Nisonoff, A., Margoliash, E., and Reichlin, M. (1969). *Science* 155:1273.
Nisonoff, A., Reichlin, M., and Margoliash, E. (1970). *J. Biol. Chem.* 245:940.
Nitecki, D. E., and Goodman, J. E. (1966). *Biochemistry* 5:665.
Nitz, R. M., Mitchell, B., Gerwing, J., and Christensen, J. (1969). *J. Immunol.* 103:319.
Noble, R. W., Reichlin, M., and Gibson, Q. H. (1969). *J. Biol. Chem.* 244:2403.
Oken, D. E., and Biber, T. V. L. (1968). *Am. J. Physiol.* 214:791.
Ovary, Z. (1964). *Immunochemistry* 1:241.
Patterson, R., Lucena, G., Metz, R., and Roberts, M. (1969). *J. Immunol.* 103:1061.
Paul, W. E., Katz, D. H., Goidl, E. A., and Benacerraf, B. (1970). *J. Exptl. Med.* 132:283.
Pelichova, H., Suzuki, T., and Cinader, B. (1970). *J. Immunol.* 104:195.
Perlstein, M. T., and Atassi, M. Z. (1971). *Fed. Proc.* 30:531.
Perutz, M. F. (1965). *J. Mol. Biol.* 13:646.
Perutz, M. F., Liquori, A. M., and Eirich, F. (1951). *Nature* 167:929.
Pope, C. G. (1966). *Adv. Immunol.* 5:209.
Pressman, D., and Grossberg, A. L. (1968). *The Structural Basis of Antibody Specificity,* W. A. Benjamin Inc., New York.
Prout, T. E. (1962). *J. Chron. Dis.* 15:879.
Rajewsky, H., Schirrmacher, V., Nase, S., and Jerne, N. K. (1969). *J. Exptl. Med.* 129:1131.
Rappaport, I. (1965). *Adv. Virus Res.* 2:223.

Rappaport, I., and Zaitlin, M. (1970). *Virology* 41:208.
Reichlin, M. (1970). *Immunochemistry* 7:15.
Reichlin, M. (1969). *Fed. Proc.* 28:435.
Reichlin, M., Hay, M., and Levine, L. (1963). *Biochemistry* 2:971.
Reichlin, M., Bucci, E., Antoni, E., Wyman, J., and Rossi-Fanelli, A. (1964a). *J. Mol. Biol.* 9:785.
Reichlin, M., Hay, M., and Levine, L. (1964b). *Immunochemistry* 1:21.
Reichlin, M., Bucci, E., Fronticelli, C., Wyman, J., Antonini, E., and Rossi-Fanelli, A. (1965a). *J. Mol. Biol.* 12:774.
Reichlin, M., Bucci, E., Wyman, J., Antonini, E., and Rossi-Fanelli, A. (1965b). *J. Mol. Biol.* 11:775.
Reichlin, M., Bucci, E., Fronticelli, C., Wyman, J., Antonini, E., Ioppolo, C., and Rossi-Fanelli, A. (1966). *J. Mol. Biol.* 17:18.
Reichlin, M., Margoliash, E., and Nisonoff, A. (1968a). *Fed. Proc.* 27:276.
Reichlin, M., Schnure, J. J., and Vance, V. K. (1968b). *Proc. Soc. Exptl. Biol. Med.* 128:347.
Reichlin, M., Nisonoff, A., and Margoliash, E. (1970). *J. Biol. Chem.* 245:947.
Richards, F. M., and Vithayathil, P. J. (1959). *J. Biol. Chem.* 234:1459.
Roelants, G. E., and Goodman, J. W. (1970). *Nature* 227:175.
Salvin, S. B., and Liauw, H. L. (1967). *Arch. Allergy* 31:366.
Scheraga, H. A. (1967). *Fed. Proc.* 26:1380.
Schlossman, S. H., and Levine, H. (1970). *Cell. Immunol.* 1:419.
Schwick, H. G. (1966). *Behringwerk-Mitleilungen* 46:87.
Seeger, R. C., and Oppenheim, J. J. (1970). *J. Exptl. Med.* 132:44.
Sela, M. (1966). Adv. *Immunol.* 5:29.
Sela, M. (1967). In Killander, J. (ed.), *Nobel Symposium 3, Gamma Globulins,* Interscience, Stockholm, pp. 455.
Sela, M. (1969). *Science* 166:1365.
Sela, M., Anfinsen, C. B., and Harrington, W. F. (1957). *Biochim. Biophys. Acta* 26:502.
Sela, M., Schechter, B., Schechter, I., and Borek, F. (1967). *Cold Spring Harbor Symp. Quant. Biol.* 32:537.
Senyk, G., Nitecki, D., and Goodman, J. W. (1971a). *Science* 171:407.
Senyk, G., Brady Williams, E., Nitecki, D. E., and Goodman, J. W. (1971b). *J. Exptl. Med.* 133:1294.
Shinka, S., Imanishi, M., Kuwahara, O., and Amano, T. (1962). *Biken J.* 5:181.
Shinka, S., Imanishi, M., Miyagawa, N., Amano, T., Inouye, M., and Tsugita, A. (1967a). *Biken J.* 10:89.
Shinka, S., Imanishi, M., Miyagawa, N., Amano, T., Inouye, M., and Tsugita, A. (1967b). *Proc. Jap. Acad.* 43:505.
Simonsen, M. (1967). *Cold Spring Harbor Symp. Quant. Biol.* 32:517.
Singer, S. J. (1965). In Neurath, H. (ed.), *The Proteins,* Vol. III, 2nd ed., Academic Press, New York, pp. 269.
Singer, S. J., and Richards, F. M. (1959). *J. Biol. Chem.* 234:2911.
Skeggs, L. T., Jr., Kahn, J. R., Lentz, K., and Shumway, N. P. (1957). *J. Exptl. Med.* 106:439.
Smyth, D. G., Stein, W. H., and Moore, S. (1963). *J. Biol. Chem.* 238:227.
Spitler, L., Benjamini, E., Young, J. D., Kaplan, H., and Fudenberg, H. H. (1970). *J. Exptl. Med.* 131:133.
Spragg, J., Austen, K. F., and Haber, E. (1966). *J. Immunol.* 96:965.
Spragg, J., Schroeder, E., Stewart, J. M., Austen, K. F., and Haber, E. (1967). *Biochemistry* 6:3933.
Spragg, J., Talamo, R. C., Suzuki, K., Appelbaum, D. M., and Austen, K. F. (1968). *Biochemistry* 7:4086.
Stavitsky, A. B. (1954). *J. Immunol.* 72:360.
Stewart, J. M., Young, J. D., Benjamini, E., Shimizu, M., and Leung, C. Y. (1966). *Biochemistry* 5:3396.

Suzuki, T., Pelichova, H., and Cinader, B. (1969). *J. Immunol.* 103:1366.
Talmage, D. W., Radovich, J., and Hemmingsen, H. (1970). *Adv. Immunol.* 12:271.
Tanaka, M., Nakashima, T., Benson, A., Nower, H., and Yasunobu, K. T. (1966). *Biochemistry* 5:1666.
Tanford, C., and Hauenstein, J. D. (1965). *Biochim. Biophys. Acta* 19:535.
Teather, R. M., and Gerwing Levy, J. (1971). In preparation.
Thompson, K., Benjamini, E., and Harris, M. (1971). In preparation.
Touber, J. L., Stoll, R. W., Ensinck, J. W., and Williams, R. H. (1970). *Diabetes* 119:409.
Tsugita, A., Gish, D., Young, J., Fraenkel-Conrat, H., Knight, C. A., and Stanley, W. M. (1960). *Proc. Natl. Acad. Sci.* 46:1463.
Unger, R. H., Eisenstraut, A. M., McCall, M. S., and Madison, L. L. (1961). *J. Clin. Invest.* 40:1280.
van Regenmortel, M. H. V. (1966). *Adv. Virus Res.* 12:207.
Van Vunakis, H., Leikhim, E., Delaney, R., Levine, L., and Brown, R. K. (1960). *J. Biol. Chem.* 235:3430.
Varandini, P. T. (1967). *Biochemistry* 6:100.
von Sengbusch, P. (1965). *Z. Verebungsl.* 96:364.
Watanabe, S., Okada, Y., and Kitagawa, M. (1967). *J. Biochem.* 62:150.
Waterfield, D., Gerwing Levy, J., and Kilburn, D. G. (1971). In preparation.
Williams, A. W., and Reichlin, M. (1968). *Fed. Proc.* 27:259.
Wilson, S. (1967). In Cinader, B. (ed.), *Antibodies to Biologically Active Molecules,* Pergamon Press, Oxford, pp. 235.
Wilson, S., Aprile, M. A., and Sasaki, L. (1967). *Can. J. Biochem.* 45:1135.
Wyckoff, H. W., Hardman, K. D., Allewell, N. M., Inagami, T., Johnson, L. M., and Richards, F. M. (1967). *J. Biol. Chem.* 242:3984.
Yagi, Y., Maier, P., and Pressman, D. (1965). *Science* 147:617.
Yagi, Y., Maier, P., and Pressman, D. (1967). *Fed. Proc.* 26:703.
Young, J. D., and Leung, C. Y. (1970). *Biochemistry* 9:2755.
Young, J. D., Benjamini, E., Shimizu, M., and Leung, C. Y. (1966). *Biochemistry* 5:1481.
Young, J. D., Benjamini, E., Stewart, J. M., and Leung, C. Y. (1967). *Biochemistry* 6:1455.
Young, J. D., Benjamini, E., and Leung, C. Y. (1968). *Biochemistry* 7:3113.

Human Histocompatibility Antigens

Ralph A. Reisfeld
Department of Experimental Pathology
Scripps Clinic and Research Foundation
La Jolla, California
and

Barry D. Kahan
Department of Surgery
Massachusetts General Hospital
Boston, Massachusetts

I. INTRODUCTION

Transplantation antigens are genetically segregating, cell surface markers which trigger an immune response by the host following the grafting of foreign tissues. The hypothesis that transplants are destroyed by an immunological reaction is based upon Medawar's demonstration of specific accelerated rejection of repeat tissue grafts derived from the original donors, i.e., the second-set phenomenon (Medawar, 1963). Allografting thus causes the host to recognize a set of cellular antigens which distinguish him from the graft donor. These polymorphic antigenic specificities constitute an allotypic system under the control of genetic loci. In most species, there is a single histocompatibility locus controlling the rapid rejection of allografts: H-2 of mice, B locus of chickens, Ag-B of rats, C-LA of chimpanzees, D-LA of dogs, and HL-A of man (Ceppellini *et al.*, 1967; Dausset *et al.*, 1970). The strong loci code for a mosaic of serological specificities. One proposal organizes the alleleic specificities into two subloci: one consisting of HL-A1,2,3,9,10,11 and the other of HL-A5,7,8,12,13 (Dausset *et al.*, 1969). Although the two-subloci concept is consistent with the serological data from family studies, it fails to explain (1) the inheritance of the widespread, broad specificities, e.g., 4a,4b,6a,6b, and (2) the production of antibodies directed either against specificities present in the host or absent from the immunizing cell (Batchelor and Selwood, 1970). Although several alternative inheritance patterns have been recently discussed, it is anticipated that the

51

genetic mechanisms will be apparent once the chemical structure of the gene-product transplantation antigens is elucidated. In spite of these theoretical uncertainties, there is a definite correlation between HL-A phenotypes and tissue graft survival. A match of the antigenic determinants of donor and host provides a better chance of transplant survival.

The function of the cell surface-located transplantation antigens is almost certainly independent of the artificial situation of transplantation immunity. Theories concerning the physiological significance of these antigens are based on their strategic location and their relatively strong immunogenicity. Although these membrane-bound substances are readily detected by their participation in the allograft reaction, they probably exert an important role in cell economy as receptor sites in immunosurveillance, growth regulation, permeability, or contact phenomena, or as an integral part of the membrane architecture. The concept of immunosurveillance proposes that histocompatibility antigens function as surface receptors allowing the immune system to distinguish autologous cells from foreign cells (Thomas, 1959; Burnet, 1970). Based upon the findings that the reaction of lymphoid cells with anti-immunoglobulin sera interferes with the development of graft vs. host reactions and with the adoptive transfer of delayed-type hypersensitivity, it has been suggested that lymphocyte messengers with immunoglobulin determinants structurally complementary to transplantation antigens might decipher the mosaics of cell surface receptors in the body (Mason and Warner, 1970).

The interesting relationship between immunological reactivity and histocompatibility phenotype has prompted Jerne to propose a general theory of the ontogeny of responsiveness. Structural genes in the germ line code for antibody molecules specifically directed against histocompatibility antigens. However, within these genes are variable cistronic regions coding for antibody molecules not directed against histocompatibility antigens. Subsequently, there is a specific suppression of nonmutant self-recognizing clones, i.e., clones destined to produce antibody against the host's own histocompatibility antigens, leaving only those cells recognizing foreign allospecificities. Each individual has a different range of antibody reactivities since he possesses only a fraction of the possible mutant antibody diversity (Jerne, 1971).

However, hypotheses supposing the function of transplantation antigens to be generation, maintenance, or expression of the immune response do not explain the retention of histocompatibility antigens by cells perpetuated in long-term cluture (Kahan and Reisfeld, 1969a). The persistence of these antigens *in vitro* suggests that they play a fundamental role in overall cell economy. An antibody response directed against these antigens inflicts serious cell damage. Through their inadvertent participation in allograft immunity, they offer one of the most accessible and specific systems for the study of cell membranes. Thus elucidation of the physiology and chemistry of histocompatibility antigens should contribute to a better understanding of cell membrane structure and function.

II. EXTRACTION OF HISTOCOMPATIBILITY ANTIGENS

A considerable body of evidence suggests that the alloantigenic determinants controlled by histocompatibility loci are arrayed on cell surface membranes. Early studies with agglutinating alloantisera indicated that the major portion of the strong alloantigenic determinants was associated with the cell surface. This location was confirmed by a number of studies with (1) fluorescent antibody (Möller, 1961; Cerottini and Brunner, 1967; Gervais, 1968); (2) antibody absorption before and after cell rupture (Haughton, 1964); and (3) purified membrane fractions (Ozer and Hoelzl-Wallach, 1967). More recently, studies employing antibody blocking reagents recognizing five murine alloantigenic systems (H-2, θ, Ly-A, Ly-B, and TL) suggest that these specificities comprise a single cluster or basic segment of the repetitive membrane structure (Boyse et al., 1968). The proximity of allotypic specificities was confirmed with ferritin-tagged antibodies for H-2 (Davis and Silverman, 1968) and HL-A (Kourilsky et al., 1971). These antigenic hot points are consistent with the membrane model of complex mosaics of lipoprotein assemblies with architectural organization which are distributed over the cell surface in discrete patches of 1.2 Å^2 (Wallach and Gordon, 1969). While there is a variety of evidence suggesting that these specificities are present on all membranous structures, it has been difficult to conclusively demonstrate them at internal sites in the cell (Kahan and Reisfeld, 1969a).

Chemical and immunochemical characterization of histocompatibility determinants in the form of membrane fragments is exceedingly difficult, if not impossible, since these fragments contain multiple antigenic systems which may be implicated in allospecificity, including strong and weak transplantation antigens as well as blood group specificities ABO, Mn, and P (Kahan and Reisfeld, 1971). While the initial efforts in the field of transplantation antigen chemistry were directed toward characterizing poorly soluble membrane pieces, it soon became clear that histocompatibility markers had to be isolated independent of the membranous network. After a decade of research, these membranous antigens were liberated and solubilized from cell surfaces, and thereby began a new era in transplantation chemistry.

The chemical problem was then defined: (1) to solubilize materials from the strongly apolar glycolipoprotein matrix of the cell membrane, of which the protein portion itself might contribute significantly to transplantation antigenicity, and (2) to purify the active principle from the overwhelming array of contaminant materials. Criteria of solubility of antigenic materials were formulated based upon the ultracentrifugal behavior, absence of membrane particulates on electron microscopy, biological behavior, and amenability to chemical purification (for review, see Kahan and Reisfeld, 1969a; Reisfeld and Kahan, 1970a). An array of different procedures was employed to achieve this goal, including use of detergent and organic solvents, sonic energy, proteolytic digestion, and treatment with simple salts. Although all of these methods release histocompatibility antigens, they are nonselective in that they also release large amounts of contaminant materials. Consequently, initial antigen extracts, although biologically potent, are chemically extremely complex. The efficient application of selective immunochemical procedures for purification of these materials will require elucidation of the physicochemical nature of these antigens.

Two strategies have been utilized based on presently available methods for the solubilization of histocompatibility antigens: (1) covalent bond cleavage and (2) disruption of noncovalent intermolecular interactions. Successful covalent cleavage resulting in antigen solubilization has been obtained with phospholipase A treatment of detergent-extracted membranous material and with proteolytic digestion of cellular membrane eluates. The products released by these methods are extremely heterogeneous in chemical and serological senses (Kandutsch, 1960, 1963).

Prolonged incubation of lymphoid tissue membrane fractions at room temperature (autolysis) released human and murine alloantigens (Nathenson and Davies, 1966), which were found to be chemically highly complex, high molecular weight, polydisperse substances following Sephadex exclusion chromatography (Davies, 1967). Careful serological analysis showed that they possessed a low degree of immunological specificity (Sanderson, 1968). The initial suggestion that autolysis solubilized up to 14% of the antigen on membrane fractions has now been amended to a yield of only 1-2% of the activity (Shimada and Nathenson, 1969). Since the autolytic method, which depends upon the action of native cathepsins, has proven difficult to reproduce, concerted proteolysis with papain has been employed to solubilize human and murine alloantigens (Davies, 1968) (enzyme/substrate ratios varying from 2:1 to 1:150; Sanderson and Batchelor, 1968; Mann et al., 1968; Shimada and Nathenson, 1969). The materials released by papain digestion have been most extensively studied and were found to be of limited utility. First, prolonged incubation of membrane fragments with papain destroys antigenic activity. Second, the yield of purified material is distressingly low. From 4000 mouse spleens calculated to contain about 40 mg of antigen, papain solubilized about 130-fold more protein than antigen present (5310 mg). Subsequent purification retrieved three electrophoretic fractions which in total contained 1.43 mg (Shimada and Nathenson,

1969). Although the specific activity of the material had increased, 98% of the total activity units present in the crude extract was lost. Furthermore, Mann *et al.* (1969) found that only 0.025 mg of antigen could be obtained following electrophoresis of papain-solubilized HL-A antigen derived from 3.8×10^9 cultured lymphoid cells. Therefore, these investigators have recommended that covalent cleaving methods be abandoned, since they found that treatment with sodium dodecyl sulfate yielded more potent antigens (Mann and Levy, 1971). Third, papain solubilizes certain specificities only poorly and apparently destroys other antigens (Shimada and Nathenson, 1969). Finally, papain as a nonspecific proteolytic enzyme cleaves not only the antigen from the membrane but also the antigenic molecule itself. Thus the choice of this enzyme as a first step in the solubilization of chemically highly complex, membrane-associated antigens is rather unfortunate as it precludes any understanding of the structure of the macromolecules on the cell surface because it probably destroys both subunit and primary structure.

On the other hand, methods not depending upon covalent bond cleavage are seemingly less destructive to antigenic structure and leave antigenic moieties relatively intact. The detergent techniques (Triton, decyl and dodecyl sulfate, deoxycholate, and sodium lauroyl sarcosinate) and the butanol extraction method yield stabilized suspensions rather than truly solubilized materials. These antigens have been obtained in low yields, in essentially a water-insoluble state, and of poor specific activity (for review, see Reisfeld and Kahan, 1970*a*).

The first truly soluble materials with transplantation antigenic activity were prepared by exposure to a controlled burst of low-intensity sound (Kahan, 1965). Biologically potent histocompatibility antigens were obtained with sonic energy (below 16,000 cycle/sec) but not with ultrasonic energy (above 16,000 cycle/sec. Ultrasonic denaturation of antigenic determinants probably is related to the destructive effects of this energy on molecular configuration, including covalent bond cleavage, and to marked local heating effects. Sonic energy is deployed by the Raytheon model DF 101 magnetostrictive oscillator, in which a plate voltage generates oscillations of a laminated rod seated within the coil excitation field. The oscillations are transduced on a diaphragm generating sonic waves in the treated fluid. The yield of solubilized antigen depends upon the conditions of sonication: the frequency, intensity, temperature, cell concentration, and exposure time. A brief exposure (3-5 min) of cell suspensions of 25-50 $\times 10^6$ cells per milliliter to diaphragm-mediated sound of 9-10 kc/sec, 15.5 w/cm^2, 4°C, liberates 12-15% of the total antigenic activity of murine, guinea pig, dog, or human cells in soluble form (Reisfeld and Kahan, 1970*a*). Although the mechanism of sonic liberation is not fully understood, it is thought to represent a depolymerization of noncovalent interactions between the antigen and insoluble membrane components (for review, see Kahan and Reisfeld, 1971).

In an effort to improve the yields of antigen and to obtain a simpler method for widespread use in the solubilization of transplantation antigens, the technique of salt extraction with 3 M potassium chloride (KCl) has been introduced and is now widely accepted (Reisfeld et al., 1970b, 1971a,b). True solubilization of antigens from the membranous network posed a formidable problem since the molecular organization of cell membranes is predominated by hydrophobic bonds. Van der Waal forces between apolar groups are weak, and hydrogen bonds of the C–O ... H–N and C–O ... H–O type are thermodynamically unstable unless protected from water (Klotz and Farnham, 1968). Thus, apolar groups form hydrophobic bonds mainly due to their thermodynamically unfavorable interaction with water. The transfer of an apolar molecule from a lipophilic environment to water is accompanied by a decrease in entropy caused mainly by the highly ordered structure of water. However, the negative entropy change can be diminished by any disorder in the structure of water. Such a decrease in the ordered structure of water is effected by physical disruption, e.g., sonic energy, or by anions, e.g., SCN^-, ClO^-_4, I^-, Br^-, Cl^-, which have large positive entropies due to their structure-breaking effects on water. In addition, these anions also decrease the polarity of the surrounding water, making it more lipophilic. Consequently, the hydrophobic attractions mainly responsible for the native structure of membranes are weakened, and the entry of the apolar residues into the aqueous phase is greatly facilitated (Hatefi and Hanstein, 1969). Because of such properties, these anions are often called chaotropic agents (Hatefi and Hanstein, 1969). These agents can dissociate antigen-antibody complexes without affecting antibody specificity (Dandliker et al., 1967) and can resolve microsomal enzymes and complexes of the mitochondrial electron transport system (Hatefi and Hanstein, 1970).

Although KCl is a weak and relatively ineffective chaotropic agent for the resolution of substances associated with intracellular membrane systems (Hatefi and Hanstein, 1970), it has proven to be highly effective for the solubilization of membrane-associated HL-A antigens from cultured lymphoid cells. In contrast to sonic exposure, extraction with 3 M KCl results in rather extensive destruction of the internal cytoarchitecture, releasing nuclear DNA and nuclear proteins. On the other hand, KCl liberates much less lipoprotein from the cell surface than does sonication (Reisfeld et al., 1971a,b). The reduction in lipoprotein is advantageous since these substances interfere with the purification of the antigenic principle by aggregation phenomena and by the trapping of active moieties. The major advantages of the KCl method are its simplicity and the reproducibility of the high yields of soluble HL-A antigens. Recently, this method has been found to be applicable to the extraction of soluble H-2 antigens from L1210 cultured murine cells, of Ag-B substances from rodent spleens, and of tumor-specific determinants from guinea pig carcinogen-induced neoplasms (Reisfeld et al., 1971a).

III. SOLUBILIZATION AND PURIFICATION OF HL-A ANTIGENS

A. Antigen Source Material

Cells and tissues from a variety of sources including man, mouse, dog, guinea pig, and rat have served as sources for the extraction of transplantation antigens by sonication or by hypertonic salt extraction (Kahan and Reisfeld, 1969a; Reisfeld and Kahan, 1971). Although inbred lines of a few species have been the key to the understanding of transplantation antigenic systems in animals, they have not provided an adequate source for the extraction of sufficient amounts of antigen to permit rigorous chemical characterization. Consequently, progress in this area has been exceedingly slow. In the case of human histocompatibility substances, the situation was even more difficult, since until quite recently there existed no large source of HL-A antigens of uniform genetic constitution. The organs of a single donor yield only very limited amounts of material, and comparisons among various preparations are not meaningful since the phenotype of each donor is different (Kahan et al., 1968).

Cultured human lymphoid cell lines derived from normal donors and from those with lymphoid malignancies were first perpetuated on a large scale by Moore et al. (1967). The numerous objections to the use of cultured lines derived from donors with malignancies for HL-A antigen extraction were overcome by the utilization of cultured cells derived from normal donors and by the treatment of these cells with low-frequency sound, detergents, and hypertonic salt to procure soluble HL-A antigens (Reisfeld et al., 1970a,b). The large cultured lymphoblasts contain quantitatively more antigen than do the donor's peripheral lymphocytes, have a normal karyotype, display neither virus particles nor EB surface antigens, and possess all of the HL-A specificities detectable on the donor's peripheral lymphocytes (Reisfeld et al., 1970a,b).

Challenging problems which have been unapproachable with inbred lines of animals are now being investigated with continuous cell cultures. For example, radiolabeled precursor amino acids can be fed to cultured cells, thus facilitating exacting microanalytical analysis of purified antigens. Pulse labeling of synchronous cell lines offers a means to an understanding of the biosynthetic pathways of HL-A antigens. Most important is the relative ease with which large amounts of cells can be grown in tissue cultures and genetically uniform lines can be maintained for long periods of time. Although at present most lymphocytic lines possess multiple antigenic determinants, it may eventually be possible through screening programs to culture lymphocytes from members of a single family, i.e., "congenic lines" that differ by a minimal number of antigenic determinants. It also seems feasible that in vitro fusion by viral agents might produce selection pressures ultimately leading to the suppression of some anti-

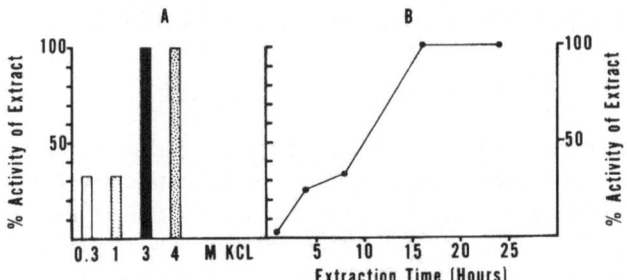

Figure 1. KCl extraction of HL-A antigens. A: Effect of KCl molarity on antigen yield. B: Effect of time of extraction on antigen yield.

genic determinants, yielding "mutant lines." The antigenic products of such lines would offer an excellent model for studying the effect of genetic expression of alloantigenic determinants on the chemical structure and configuration of HL-A antigens, which may lead to an understanding of the chemical basis of the compelling problems in human transplantation individuality.

B. KCl Extraction Technique

During a search for effective methods to solubilize HL-A antigens, we found that extraction of cultured lymphoid cells with hypertonic salt solution (0.3 M KCl) yielded reasonable amounts of soluble antigens (Reisfeld and Kahan, 1970b). Subsequently, an extensive effort was made to develop a simple, reproducible hypertonic salt extraction method which could yield large amounts of soluble HL-A antigens. Experiments designed to determine optimal extraction conditions revealed that 3 M KCl solubilized more antigen than either 0.3 or 1.0 M and that a 16 h extraction time gave maximal yields. In addition, we observed that solubilization was most efficient at 4°C and that at 40°C there was a considerably lower antigen yield (Fig. 1) (Reisfeld et al., 1970b, 1971a,b).

In order to extract soluble HL-A antigens, up to 50×10^9 dispersed cultured cells are suspended in 0.9% sodium chloride containing 3 M KCl buffered at pH 7.4 (10-20 ml solvent per 10^9 cells). This suspension is gently agitated on a mechanical shaker for 16 hrs at 4°C and then centrifuged at 163,000 x g for 1 hr. Subsequently, the extract is dialyzed against large volumes of saline, and a gelatinous precipitate forms which contains primarily DNA. This material, which apparently results from lysis of nuclei by 3 M KCl, is removed by centrifugation at 1500 x g for 20 min. DNA could not be detected in this supernatant material either by the diphenylamine assay of a hot trichloroacetic acid extract (Burton, 1956) or by radiolabeling experiments. For example, the antigen in the 1500 x g supernatant fluid was from two generations of cultured

WIL$_2$ lymphoid cells which had been uniformly labeled with thymidine-2-[14]C (0.1 μl per 5 x 10^5 cells); this antigen contained less than 1% of the total labeled DNA (Reisfeld *et al.*, 1971*a*).

The viability of cultured cells is important in order to obtain maximal antigen yields, as a batch of cells with 23% viability yielded only 9% soluble antigen with poor immunological potency (specificity ratio 18). Furthermore, only small yields (3-8%) of soluble HL-A antigen could be obtained by KCl treatment of crude cell membrane preparations (Reisfeld *et al.*, 1971*a*) obtained either by freeze-drying and hypotonic salt extraction (Mann *et al.*, 1969) or by nitrogen bomb decompression (Hunter and Commerford, 1961) of cultured lymphoid cells.

C. Purification with Polyacrylamide Gel Electrophoresis

From a large body of experience with soluble serum proteins, it would seem rather logical to employ as many different fractionation principles as possible for the purification of histocompatibility antigens, including fractional salting-out, exclusion chromatography, ion exchange chromatography, and electrophoresis. However, our experience with these antigens revealed that their biological activity is lost by extensive manipulation. Indeed, this has been confirmed by the observations of a number of workers (Mann *et al.*, 1968, 1969). Following an extensive multistage protocol, up to 98% of the biological activity present in crude papain-solubilized murine histocompatibility antigen extracts was lost, and the product was still not electrophoretically homogeneous (Shimada and Nathenson, 1969). Consequently, concepts derived from the purification of soluble serum proteins do not necessarily apply to histocompatibility antigens, which are rendered soluble with considerable difficulty due to their membranous association with hydrophobic and insoluble substances. Indeed, the "solubility" of these antigens may be in precarious balance, and a number of factors might render them less soluble at any time: (1) sudden removal of other components in a mixture or (2) changes in their local environment resulting from absorption-desorption or alteration in ionic strength and/or charge during chromatography. The best strategy thus seems to be the application of a single procedure which affords maximal resolution. This criterion is met by preparative acrylamide gel electrophoresis in a discontinuous buffer system, a method which at present proves to be the procedure of choice for the purification of histocompatibility antigens.

Polyacrylamide gel electrophoresis (PAGE), because of its high resolving power, sensitivity, and versatility, has found a very wide application since its introduction (Raymond and Weintraub, 1959; Ornstein, 1964). Although this method has been used mainly as an analytical tool, with qualitative patterns being utilized for the interpretation of results, mathematical parameters of the physicochemical nature of PAGE have been described recently (Rodbard and Chrambach, 1971). The method is highly efficient, as zone electrophoresis in polyacrylamide gels simultaneously exploits differences in size and charge.

Quantitative PAGE is based upon (1) the achievement of a high degree of pore reproducibility (over the range of 0.5–3 nm) by a simple adjustment of the total concentration of acrylamide and its cross-linking reagent and (2) the development of a theory for the movement of molecules through gels, yielding a computer program which generated up to 4000 discontinuous buffer systems operative between pH 3 and 11. Thus, the charge separation between two molecules can be maximized when the net charges of the molecule of interest and the contaminant(s) are known. This general approach is based upon a statement of the electrophetric data in "Ferguson plots," i.e., the log of mobility vs. gel concentrations, which yield via a computer methodology information on molecular size, free mobility, and valence of protein molecules (Rodbard and Chrambach, 1971). Of considerable interest is the correspondence between the molecular size of KCl-solubilized antigen, estimated by the retardation co-efficient calculated by this technique to be 32,000, and the molecular weight of this same purified antigen preparation, determined to be 31,000 by ultra-centrifugal analyses (Reisfeld, unpublished observations). Electrophoretically purified HL-A antigen solubilized by low-frequency sound had a molecular weight of 34,600 as determined by ultracentrifugal analyses (Kahan and Reis-feld, 1969c. Numerous preparative electrophoreses utilizing varying pore size conditions have shown clearly that the active antigenic moiety and neighboring contaminants could be resolved optimally at a gel concentration of 7½ % T (percent acrylamide plus methylenebisacrylamide) and 2% C (percent methy-lenebisacrylamide divided by T). Attention to the variable of pore size has thus permitted optimal fractionation of biologically active histocompatibility antigen and has yielded physical values which are consistant with those previously determined by other methods. Electrophoretic resolution of HL-A antigen was evaluated at several different pH values (3.56, 6.96, 7.35, 7.67, and 9.60). Thus far, pH 9.6 at 0°C seems to be optimal for effective resolution (Reisfeld, unpublished observations).

Preparative PAGE has been successfully used to isolate HL-A antigens from crude antigenic extracts obtained either by application of low-frequency sound or by KCl extraction. In the case of KCl-solubilized HL-A antigens, the 1500 × g supernatant material is dialyzed against pH 6.7 tris-phosphate buffer (0.045 tris, 0.032 M H_3PO_4). As much as 100 mg protein can be applied to a Buchler Polyprep 100 column. Protein loads as high as 200 mg can readily be applied to the Buchler Polyprep 200 column, which has recently come into increased use in our laboratory. Electrophoresis is performed at 0°C in system "B" (Rodbard and Chrambach, 1971), pH 9.6, 7½% acrylamide gel at a constant current of 35 mA. Fractions (8 ml) are collected at a flow rate of 0.8 ml/min

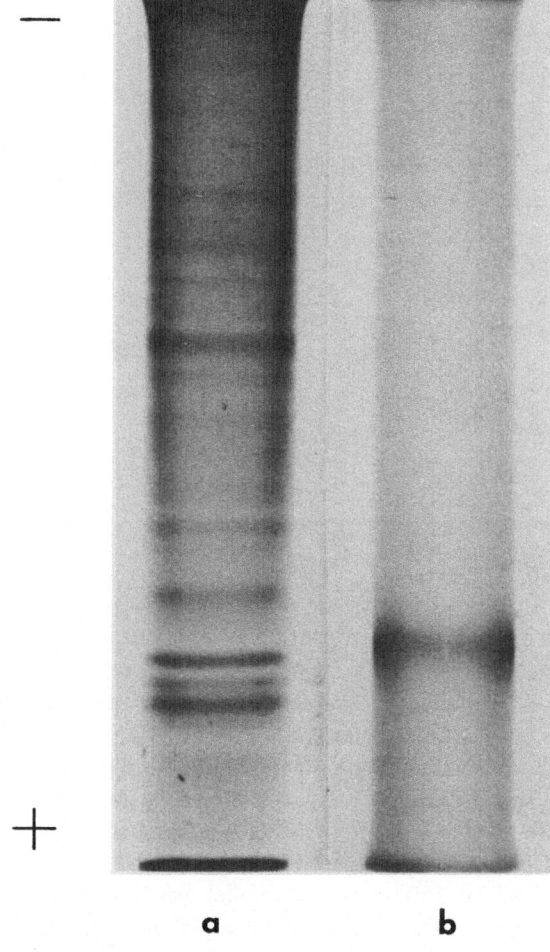

Figure 2. Electrophoretic patterns of HL-A antigens. (a) KCl extract prior to preparative acrylamide gel electrophoresis. (b) Antigen moiety isolated by preparative electrophoresis.

with a tris-HCl elution buffer (0.138 M tris, 0.18 M HCl, pH 8.2, containing 10% v/v sucrose). The fraction eluting at R_f 0.75–0.78 exhibits specific antigenic activity, although at times some activity is detected in the upper gel, possibly due to aggregation of the antigenic material. Electrophoretic moieties comprising R_f 0.75–0.80 are re-electrophoresed at pH 9.6, and the fraction of R_f 0.78 contains the bulk of the antigenic activity. Upon re-electrophoresis of this fraction in an analytical system (Fig. 2) (7½ % acrylamide gel, 0°C, pH 9.6), it is

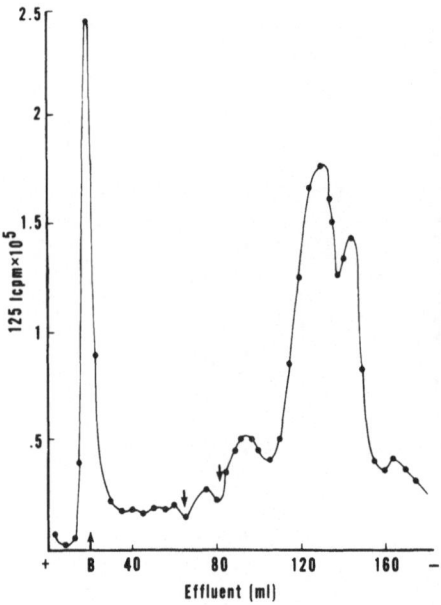

Figure 3. Elution diagram of preparative acrylamide gel electrophoresis of
KCl extract of HL-A antigen (RPMI
1788, 100 mg). Arrows denote the region of antigenic activity.

found to consist of a single electrophetric component, as has been shown with
the HL-A antigens solubilized by sonication (Reisfeld and Kahan, 1970*b*).

In order to obtain an accurate assessment of the antigenic activity of
purified HL-A antigens, minute amounts of proteins have to be determined. For
this purpose, a radiolabel technique is combined with Kjeldahl nitrogen analysis.
For example, 1 billion cultured cells in log growth phase (2×10^7 ml) were
incubated for 4 hr with a mixture of ^{3}H-labeled amino acids (2.5 mCi total) in
Eagle's minimal essential medium containing 1% of its normal amino acid
content. Approximately 15% of the radiolabel was incorporated into the cells,
while 2% was found in the ultracentrifugal supernatant fluid and 0.04% in the

Table I. Electrophoretic Purification of HL-A Antigen (WIL$_2$)

Antigen	mg N/10^9 cells	ID$_{50}$ units [a]/mg N	Total ID$_{50}$ units
Prior to electrophoresis	3.89	41,666	162,080
R_f0.78 fraction	0.089	833,333	74,166

[a] The reciprocal of the soluble protein antigen dose which inhibits cytotoxic antisera
(TO 11.03, anti-HL-A2) at zero cytotoxic units to 50% of the untreated level. This is
calculated from the ID$_{50}$ dose; e.g., if ID$_{50}$ is 0.1 μg/μ1, the ID$_{50}$ units per milligram of
protein will be 1000/0.1, or 10,000.

purified antigen. Table I shows the yield in activity of acrylamide-purified HL-A antigen. Approximately 2% of the nitrogen and 45–60% of the total antigenic activity applied to the gel column can be recovered in the antigenic component with a 20-fold increase in ID_{50} units (see below) per milligram nitrogen. Figure 3 shows a typical elution diagram obtained following preparative acrylamide gel electrophoresis of radiolabeled HL-A antigenic extracts. At present, it is possible to fractionate 300 mg of crude antigen daily while utilizing Polyprep 100 and 200 columns simultaneously, thus resolving 3-6 mg of highly purified HL-A antigen.

IV. BIOLOGICAL EVALUATION OF SOLUBLE HL-A ANTIGENS

Transplantation antigens can be detected by immune reactions which specifically recognize the allotypic distribution of these determinants. Assay systems are based upon three biological activities: graft compatibility, delayed-type hypersensitivity reactions (DTH), and the induction of and interaction with humoral alloantibody. Antigens liberated by sound have been found to mediate all of the biological activities of which intact cells are capable. Murine and guinea pig sonicated soluble antigens induce the accelerated rejection of donor-type skin grafts. Dog and rat crude sonicated suspensions induce a state such that donor-type renal allografts are accepted for a prolonged period, in some cases for more than 1 year. Guinea pig transplantation antigens mediate several DTH reactions: (1) elicitation of specific hypersensitivity cutaneous responses in allogeneic hosts presensitized with donor-type skin grafts; (2) interaction with sensitized peripheral lymphocytes in third-party local passive transfer reactions in syngeneic hosts, and with sensitized lymph node cells in nodules in the dorsal skin of irradiated hamsters; and (3) stimulation of blastic transformation of peripheral lymphocytes derived from donors sensitized with donor allografts. Murine sonicated antigen stimulates the production of specific H-2 alloantibody and interacts with that antibody *in vitro*. Human antigen elicits Arthus cutaneous hypersensitivity reactions in donors possessing specific circulating alloantibodies and inhibits the cytotoxic reactions of these alloantibodies *in vitro* (for reviews, see Kahan and Reisfeld, 1969a, 1971; Reisfeld and Kahan, 1970a).

Due to the cumbersome and time-consuming nature of compatibility studies and pending the completion of suitable toxicity studies, work on soluble HL-A antigens has been restricted to *in vitro* tests which are presumed to reflect the same biological activity that mediates histocompatibility. The two *in vitro* methods in present use are (1) inhibition of the cytotoxic reactions of specific alloantisera and (2) induction of blastic transformation of allogeneic lymphocytes. Lymphocyte transformation as evidenced by increased DNA and RNA synthesis and cell division, monitored by pulse labeling with thymidine-^{3}H to

estimate the rate of DNA synthesis, seems to reflect the recognition of histoin-
compatibility between the two admixed cell populations. Indeed, the degrees of
lymphocyte activation among members of a single family fall into separable
classes suggestive, according to statistical·analysis, of a distinct genetic mechanism (Alling and Kahan, 1969). In a similar fashion, soluble, millipore-filtered
HL-A antigens extracted with sonication or with 3 M KCl stimulate a five- to
40-fold increase in thymidine uptake by cultured histoincompatible peripheral
lymphocytes but do not affect autologous peripheral leukocytes from the donor
of the cell line (Oppenheim and Reisfeld, unpublished observations).

   Although transplantation immunity appears to depend upon a cellular
effector mechanism, humoral antibody is produced by a number of species
following allografting. Following skin grafting or the intradermal injection of
leukocytes, cytotoxic, agglutinating, and lytic antibodies are produced which
react with histocompatibility determinants. In animals, the specific alloantibodies reflect distinguishable antigenic determinants within the mosaic of the
major histocompatibility locus (Kahan and Reisfeld, 1968b; Kahan et al., 1967).
Assuming that a similar situation exists for the HL-A system, the histocompatibility activity of tissue extracts can be assayed by their capacity to specifically
block the cytotoxic reactions of typing alloantisera in vitro. However, it has not
yet been ascertained whether soluble HL-A antigens are indeed true transplantation antigens capable of affecting the survival of donor-type skin grafts.

   The microdroplet technique (Terasaki and McClelland, 1969) has been
modified to quantitate the inhibition of cytotoxic alloantibody by soluble
histocompatibility antigens (Kahan et al., 1968). Alloantigens dissolved in 1 μl
of 0.9% sodium chloride (pH 7.4) are incubated for 1 hr at 25°C with 1 μl of
alloantibody at 0.2 or 4 cytotoxic units. Then 3000 peripheral lymphocytes as
target cells and 3 μl of undiluted rabbit complement are added to the micro-
droplets protected against evaporation under oil. After an additional 4 hr
incubation period, the reactions are terminated with 1 μl of 36% formalin. The
viability of the target cell population is assessed by inverted phase contrast
microscopy. Dead cells are distinguished from live ones by their prominent
nucleus, containing clumped chromatin. A decreased potency of antibody after
preincubation with antigen is detected as an increased percentage of viable cells.

   Arithmetic plots of antigen dosage vs. percent viability of target cells are
sigmoidal in character. Since the degree of antigen inhibition is inversely proportional to the number of cytotoxic units of antibody employed, the sigmoidal curves may be converted to linear functions by the Reif (1966) modification
of the van Krogh equation:

$$\log g = \log G + \frac{1}{M} \log \left[\frac{(100 - P)}{P} \right]$$

In this equation, g is the ratio of the total quantity of antigen to that of

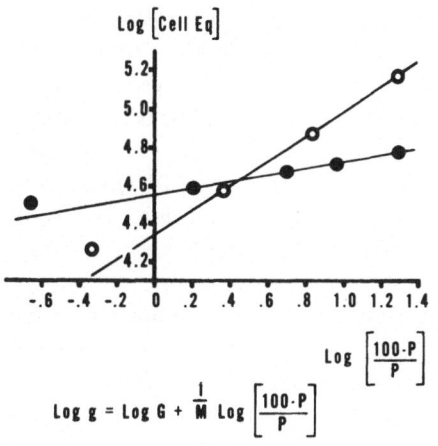

Figure 4. Van Krogh plot: HL-A2 alloantisera (TO 11.03 ○ and Pinquette ●) were incubated with serial dilutions of soluble antigen RPMI 1788 (expressed as log cell equivalents) and then tested against the same target cells. TO 11.03 had a greater sensitivity in the inhibition assay, a lower ID_{50}, and an avidity coefficient severalfold greater than antiserum Pinquette.

antiserum; P is the potency of the antiserum remaining free in the supernatant material after absorption, i.e., the number of dead cells; G equals g when P is 50%; and $1/M$ is the slope coefficient. This transformation yields several important parameters. The y intercept (ordinate) of the van Krogh plot (log G) is the 50% inhibition dose (ID_{50}), i.e., the dose required to reduce the cytotoxic effect of the treated antiserum to half that of sera not preincubated with antigen. The avidity (slope, $1/M$ coefficient) reflects the association constant of the antigen-antibody complex. A low value of $1/M$ indicates sharp absorption (Fig. 4). The ID_{50} value decreases approximately 25-fold and the avidity coefficient increases 6400-fold upon purification of the ultracentrifugal supernatant fluid of the sonicated cells to the electrophoretically homogeneous acrylamide component (Reisfeld and Kahan, 1970b). The ID_{50} values for KCl extracts derived from cell lines

Table II. Solubilization of HL-A Antigens from Cultured Lymphoid Cells

Cell line	HL-A phenotype	ID_{50}[a]				Specificity ratio[b]
		2	3	7	5	
RPMI 7249	1,2,7,8	0.1	>10	1.0	>10	>100
RPMI 4098	3	1.3	0.01	1.3	1.3	130
RPMI 1788	2,7,10, Maki	0.025	n.d.	0.09	2.5	100
WIL$_2$	1,2,5	0.016	n.d.	4.8	0.047	300

[a] Inhibition dose, ID_{50}, represents the dose of antigen required to halve the cytotoxic power of a specific alloantibody.

[b] Ratio of the concentration of antigen required to inhibit an indifferent vs. that required for a homologous antiserum. Specificity ratios were calculated for HL-A2 (RPMI 7249, RMPI 1788, and WIL$_2$) and for HL-A3 (RPMI 4098), respectively.

WIL$_2$ or RPMl 4092 were as high as 200,000 units/mg protein for the anti-HL-A serum TO 11.03 and 100,000 units/mg for the anti-HL-A3 serum Storm, respectively (Reisfeld and Kahan, 1971). A summary of HL-A antigen solubilization achieved by KCl extraction of four different cultured cell lines is shown in Table II. It is evident that antigen preparations were obtained with excellent activity in the lymphocytotoxic test; i.e., dose levels as low as 0.016 $\mu g/\mu l$ were sufficient to block the activity of specific cytotoxic alloantisera. The good immunological potency of these soluble antigen preparations is evident from their high specificity ratios.

Although the ID$_{50}$ and avidity coefficients are excellent parameters of the inhibitory activity, the immunological potency of soluble antigens is more directly related to the specificity ratio (SR) of the materials. The SR is defined as the concentration of antigen required to inhibit an antiserum directed against an antigenic determinant not present in the antigen donor vs. the concentration required to inhibit an antiserum directed against a determinant present in the donor. The ratio expresses the specific inhibitory power of an alloantigenic preparation against the homologous antiserum as opposed to an indifferent antiserum. The specificity ratios of soluble alloantigens prepared by sonic exposure or KCl exposure range from 50 to 300 depending on the nature of the preparation, the alloantisera employed, and the target cell used (Reisfeld *et al.*, 1971*b*; Kahan *et al.*, 1971). In a complementary fashion, utilizing a battery of soluble alloantigens, the specificity of a range of antisera for those soluble determinants can be quantitatively expressed as a specificity ratio. These ratios vary from about 10 to 150 in a pattern similar to that of the potency and specificity of these reagents (Reisfeld *et al.*, 1970*b*).

To determine the yield of cellular antigen in soluble form, alloantibodies were absorbed with serial numbers of intact donor cells and then tested against standard target cells (Reisfeld *et al.*, 1970*a;* Kahan *et al.*, 1971). The number of intact cells required to inhibit a specific alloantiserum by 50% was defined as the absorption dose (AD$_{50}$). The ratio of the ID$_{50}$ (expressed in cell equivalents) to the AD$_{50}$ yields the percent recovery of soluble antigen. The same antisera at the same concentration are employed in both sets of inhibition tests. Quantitative absorption studies (Table III) with soluble antigen derived from cell lines RPMI 4098, RMPI 1788, and WIL$_2$ demonstrate excellent recoveries of HL-A2,3,5,7 determinants, ranging generally from 64 to 100%. Much lower recoveries of HL-A2,7 determinants were achieved with cell line RPMI 7249. At present, there are no ready explanations for this observation. However, it should be pointed out that this cell line has been employed only infrequently for antigen extraction, whereas the other lines have been continuously employed for antigen isolation during a period of 1 year.

Although quantitative parameters of antigen activity have been defined in the serological system, the microdroplet technique is difficult and tedious,

Table III. Soluble Antigen Yields from Cultured Lymphoid Cells

Cell source	HL-A	ID_{50} units/10^9 cells	AD_{50} units/10^9 cells	Percent recovery[a]
RPMI 7249	2	350,000	1,250,000	28
	7	35,000	285,000	12
RPMI 4098	3	1,666,666	1,666,666	100
RPMI 1788	2	920,000	1,082,350	85
	7	341,000	532,800	64
WIL_2	2	2,785,712	3,061,222	91
	5	793,912	980,138	81

[a] Percent recovery = 100 × (ID_{50} units/10^9 cells) (AD_{50} units/10^9 cells).

especially since thousands of reaction droplets must be used to accurately assess the potency and immunological specificity of a single antigenic preparation. Techniques employing radiolabeled standard alloantibody are presently under development and are expected to increase the sensitivity and flexibility of the serological methods to characterize pure histocompatibility antigens.

V. CHEMICAL CHARACTERIZATION OF HL-A ANTIGENS

The approaches to characterizing HL-A antigens stem from work in animal model systems. Water-soluble transplantation antigen liberated from dispersed guinea pig cells by a brief, controlled exposure to low-intensity sound was purified to an electrophoretically homogeneous component (R_f 0.73-0.74) (Kahan and Reisfeld, 1967). The molecular weight of guinea pig transplantation antigen was determined to be 15,000 by three independent methods: (1) ultracentrifugation, employing the Yphantis sedimentation equilibrium technique; (2) exclusion chromatography on Sephadex G-200 in the presence and in the absence of 5 M guanidine-HCl; and (3) calculation from the amino acid composition (Kahan and Reisfeld, 1969c). There were characteristic, reproducible, and statistically significant differences in the amino acid compositions of antigens isolated from two inbred histoincompatible strains of guinea pigs (Kahan and Reisfeld, 1968a). The hypothesis suggesting that these animals possess transplantation allotypic specificities related to protein structure was strengthened when these antigens were found not to contain either detectable lipids or carbohydrate.

A. Antigens Prepared from Spleen Cells

In the initial work on soluble HL-A antigens, spleen cells were exposed to a brief burst of low-intensity sonic energy, essentially as described for the release of murine and guinea pig transplantation antigens. Following ultracentrifugation at 130,000 × g, exclusion chromatography on Sephadex G-200 yielded HL-A alloantigens which specifically inhibited alloantisera. Purification by semipreparative acrylamide gel electrophoresis resulted in the isolation of an electrophoretically homogeneous alloantigen (R_f 0.78-0.80). Ultracentrifugal analysis by the Yphantis sedimentation equilibrium technique revealed the antigen to be 94% monodisperse with a molecular weight of 34,600, which was in good agreement with the calculated molecular weight of 33,000 from the amino acid composition (Kahan and Reisfeld, 1969c). Electrophoretically purified HL-A antigen had two biological activities. It elicited cutaneous hypersensitivity reactions only in individuals who were actively producing monospecific cytotoxic antibodies directed against determinants present on the peripheral lymphocytes of the spleen donor. Second, it specifically inhibited 11 cytotoxic alloantisera in a pattern consistent with both the direct and the absorption HL-A typing for determinants HL-A1,2,3,5,7,8,9,12 (Kahan et al., 1968).

B. HL-A Antigens from Lymphocytes in Culture

Low-intensity sound was employed to solubilize alloantigens from the surface of lymphocytes derived from a normal human donor and perpetuated in suspension culture. The supernatant fluid following ultracentrifugation of the sonicated suspension at 130,000 × g (1 hr) inhibited the cytotoxic reactions of 21 alloantisera, reflecting 16 distinguishable HL-A antigenic determinants in a pattern identical with the direct and absorption typing of the RPMI 1788 cell line donor's peripheral lymphocytes and his cultured cells (HL-A-1-,2+,3-,5-,7+,8-,9-,10-,11-,12-,13-,Maki+). Purified soluble HL-A antigens were then isolated from a number of cultured cell lines derived from normal donors bearing different mosaics of these determinants (Reisfeld et al., 1970a). As had been previously found in the guinea pig system, the purified materials did not possess detectable lipid or carbohydrate moieties. Again, there were consistent and reproducible differences in the amino acid compositions of antigens derived from cell lines differing in their HL-A phenotypes (Reisfeld and Kahan, 1971).

Although the materials solubilized from cultured lymphocytes with sonic energy possessed marked stability (several months at -20°C) and potent serological inhibitory activity, there were some disadvantages of the sonic technique for this particular source material. First, the yield of antigen was at best 22% of the total absorptive activity on the cell membrane surface. Second, there was an excessive amount of lipoprotein solubilized from the lymphoblasts as compared to all previous tissue sources. Finally, the sonic method was rather inflexible to scale to the increased demands of preparative techniques; the 30g (3×10^{10} cells) batches of cells required division and separate sonic treatment of over 100 aliquots.

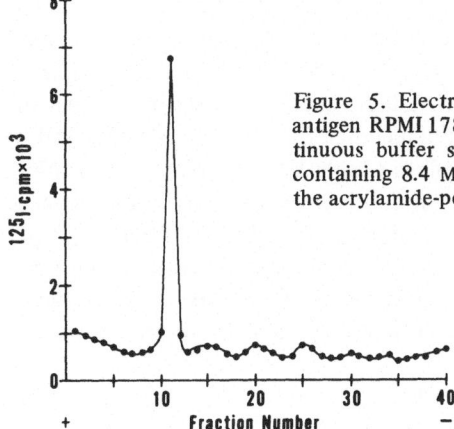

Figure 5. Electrophoretic pattern of purified ^{125}I-labeled antigen RPMI 1788. Electrophoresis was performed in a discontinuous buffer system at pH 9.6 in a 7.5% acrylamide gel containing 8.4 M urea. The lower mobility (R_f 0.72) is due to the acrylamide-pore-narrowing effect of urea.

After experimentation with a number of methods to noncovalently release HL-A antigens from cell surfaces, the 3 M KCl technique was introduced. Following 16 hr exposure of cultured cells to 3 M KCl, the supernatant material is dialyzed and then fractionated directly on preparative acrylamide gel, yielding an antigenic fraction R_f 0.75-0.78, pH 9.6, 7½% acrylamide gel (Reisfeld *et al.*, 1971*a*). Furthermore, the purified component resolves as a single moiety when subjected to electrophoresis under a variety of conditions: in the presence of either 8 M urea (Fig. 5) or 0.1% sodium dodecyl sulfate (Fig. 6), and in gels of varying porosity. Amino acid compositional analyses were quite similar to those derived from materials solubilized by sonic exposure (Reisfeld and Kahan, 1971).

Figure 6. Electrophoretic pattern of purified ^{125}I-labeled antigen RPMI 1788. Electrophoresis was carried out in 5% gel, in a continuous buffer system at pH 7.4, containing 0.1% sodium dodecyl sulfate.

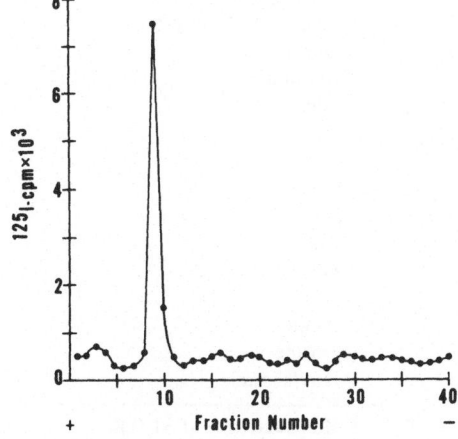

Archibald sedimentation-equilibrium analysis of purified antigen RPMI 1788 in the analytical ultracentrifuge indicated a sedimentation rate of 2.3S and a molecular weight of 31,000. Amino acid analysis of performic acid-oxidized HL-A antigen revealing four cysteic acid residues was confirmed by analysis of completely reduced carboxyamidomethylated material (in 7 M guanidine-HCl) showing 3.6 carboxymethylcysteine residues (Table IV). In order to determine whether the two disulfide bridges were inter- or intrachain, the reduced, carboxamidomethylated antigen was examined in the analytical ultracentrifuge. The reduced material had essentially the same sedimentation rate as the nonreduced material. Thus KCl-solubilized HL-A antigen seems to be a single polypeptide chain with two intrachain disulfide bridges and a molecular weight of 31,000.

Peptide maps of tryptic digests (enzyme/substrate ratio 1:50, w/w, 5 hr, 37°C) of purified HL-A antigen from cell line RPMI 1788, which had been pulse-labeled with a mixture of ^{14}C-amino acids in log growth phase, revealed a number of peptides equivalent to the number of lysine and arginine residues present in this molecule (Fig. 7). Comparison of the digests obtained from molecules differing in their HL-A phenotypes revealed a number of peptides which appeared to be distinctive for each mosaic (Reisfeld, unpublished observations). Isolation and chemical characterization of these different peptides is

Table IV. Amino Acid Composition of HL-A Antigen (RPMI 1788)

Amino acid	Residues per mole[a]
Lysine	18
Histidine	2
Arginine	5
Cysteic acid	4
S-carborymethylcrysteine	4
Aspartic acid	40
Threonine	12
Serine	18
Glutamic acid	58
Proline	12
Glycine	21
Alanine	22
Valine	11
Methionine	4
Isoleucine	8
Leucine	16
Tyrosine	4
Phenylalanine	7

[a] Molecular weight 31,000.

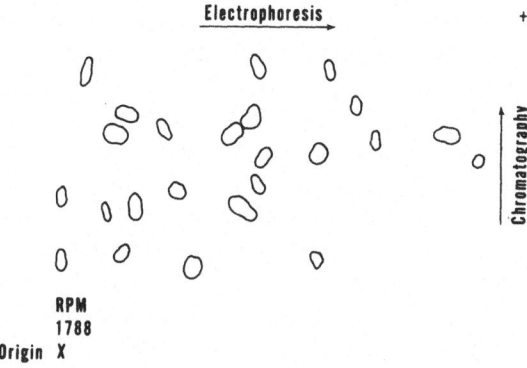

Figure 7. Tryptic peptide map of purified ^{14}C-labeled antigen RPMI 1788. Chromatography and high-voltage electrophoresis were carried out at pH 3.5.

under way and may afford the key to understanding the chemical basis of transplantation individuality.

VI. CHEMICAL NATURE OF HL-A ALLOANTIGENS

During the past 15 years, there have been numerous efforts to elucidate the chemical nature of histocompatibility antigens. Progress in this field has been exceedingly slow due to (1) the hydrophobic nature of the antigens, (2) their extreme chemical complexity, and (3) the very limited amounts of purified, soluble antigens available from genetically uniform source materials. It is not surprising that the chemical nature of the antigenic principle has been proposed to be based on the attributes of virtually every known chemical entity including DNA, lipids, carbohydrate, and protein or combinations thereof. Recent progress suggests that these antigenic determinants are essentially protein; a large body of evidence suggests that polypeptides are essential for antigenic activity (see below). On the other hand, there has been little experimental evidence which would suggest that carbohydrate moieties are essential for the expression of antigenic activity. A recent report indicates that 80% of the galactose and 25% of the N-acetylgalactosamine residues could be enzymatically cleaved from papain-treated H-2 antigens without affecting antigenic activity (Muramatsu and Nathenson, 1971). Similarly, the activity of highly purified rat tumor-specific antigens, which contain 20% carbohydrate confined to the carboxy terminal portion of the molecule, is resistant to the enzymatic cleavage of this portion of the molecule (Caputo, personal communication). Thus it seems that in these situations carbohydrate has functions other than contributing to the antigenicity of the molecules. Work by another group of investigators has implicated carbohydrate in HL-A antigenic activity. Pronase digests of papain-treated HL-A antigens were analyzed for serological inhibition of alloantisera and for their amino acid and carbohydrate compositions. The reported data from the inhibition assays were fragmentary. There appeared to be but 8-20% of the inhibitory activity recovered in glycopeptides, as determined by a 20% reduction of 75% lysis of control cells, i.e., a net inhibition of only 15% (Sanderson *et al.*, 1971). More rigorous criteria have been proposed to prove the capacity of an antigenic preparation to block cytotoxic alloantisera: 50% inhibition of 95% lysis of control cells, or a net inhibition of 47.5% (Reisfeld *et al.*, 1970*a,b*).

A large body of evidence suggests that polypeptides are essential for antigenic activity. First, histocompatibility antigens are irreversibly denatured by proteolytic enzymes, protein denaturants, low salt concentrations, *p*H values below 5 and above 10, exposure to greater than 50°C, and concentrations of detergents or complex salts which affect protein conformation (for reviews, see Kahan and Reisfeld, 1969*a*; Reisfeld and Kahan, 1970*a*). Second, purified transplantation antigens liberated by low-frequency sound contain neither lipid nor carbohydrate moieties at the detectable level (1%). Consequently, in the case of guinea pig antigen (molecular weight 15,000) there is at most one carbohydrate residue per molecule; and in the case of HL-A antigen (molecular weight 31,000) there are at most two residues of carbohydrate per molecule. Thus it is difficult to imagine that moieties other than amino acids make up the vast majority of antigenic determinants of histocompatibility antigens. Finally, amino acid analysis of electrophoretically homogeneous transplantation antigens from two histo-incompatible strains of guinea pigs and from two cell lines differing in their HL-A phenotypes showed marked and reproducible differences in their amino acid compositions (Kahan and Reisfeld, 1968*a;* Reisfeld and Kahan, 1971). The finding of marked differences strengthens the hypothesis that transplantation antigens possess genetically segregating allotypic specificities related to protein structure.

Genetic theory also supports the proposition that histocompatibility loci code for polypeptide chains rather than carbohydrate moieties as seen in the ABO system. The sharp insertions, which are noted in the ABO system and thought to reflect the actions of enzymes performing specific manuevers on precursor molecules, are not characteristic of transplantation specificities, which are frequently broad, similar to those observed with protein polymorphic systems. Furthermore, it has been suggested (Ceppellini *et al.*, 1967) that the great number of mutants which characterize the HL-A system are similar to immunoglobulin allotypes, where many mutants are distributed along a limited number of linked structural genes corresponding to many antigenic determinants of a few polypeptide chains, rather than similar to the ABO system, in which a limited number of antigenic variants are controlled by a few enzymes acting on a common substrate.

The nature of the chemical expression of the antigenic determinants on polypeptide chains is thus far unclear. Although the hypothesis suggesting that alloantigenic determinants of each of the two closely linked chromosomal regions (subloci) are present on different molecules is consistent with the serological findings of Boyse *et al.* that the $H\text{-}2^d$ and $H\text{-}2^k$ alloantigens are separate on the mouse thymocyte membrane, there is no precise chemical information to support it. Data from exclusion as well as ion exchange chromatography of papain-solubilized H-2 and HL-A antigens indicated a separation of antigenic determinants in peaks of broad, overlapping, crude fractions. However, it is impossible to assess the significance of these findings since (1) the respective peak fractions allegedly enriched in single antigenic determinants have been neither isolated nor chemically characterized; (2) the serological data have been described solely in qualitative terms; and (3) there are no specificity ratios of the fragments, which would support the claim that the relative enrichment was immunologically specific. Although it is possible that separate molecular products are determined by each of the proposed subloci, further data will be required to confirm this suggestion. HL-A2 and HL-A7 determinants of cell line RPMI 1788 and HL-A2 and HL-A5 determinants of cell line WIL_2 solubilized from cultured lymphoid cells either by low-intensity sound or by 3 M KCl extraction have not yet been separable into moieties bearing different alloantigenic specificities under a variety of electrophoretic conditions. All of the specificities in each mosaic were found in a single electrophoretic moiety. The finding that the HL-A molecule is a single polypeptide chain is compatible with either of the following: (1) more than one antigenic determinant is located on the same polypeptide chain, or (2) alloantigenic determinants controlled at different subloci are on different individual chains which are chemically, although not electrophoretically, distinct. The present data preclude a molecular model of two chains covalently linked by interchain disulfide bridges, with each polypeptide chain expressing a specificity controlled by one of the two subloci. It is tempting to propose that these chains contain regions of constant amino acid composition, probably related to the biological role of these materials in cytoarchitecture or in membrane function, which are phylogenetically homologous. Small regions of sequence variability within an otherwise constant sequence could account for the immunological specificity. The recent findings of a number of different peptides among maps of tryptic digests derived from 3 M KCl solubilized antigens of cell lines RPMI 1788 (HL-A2,7,10, Maki) and WIL_2 (HL-A1,2,5) may reflect a chemical expression of histocompatibility differences. It is now apparent that the abundant source materials available from cultured lymphoid cell lines and new, effective techniques for the solubilization and purification of relatively large amounts of homogeneous materials should permit elucidation of the structure of these chemical markers of transplantation individuality.

VII. SUMMARY

Soluble human histocompatibility (HL-A) antigens can be extracted from cultured lymphoid cell lines by a simple yet highly efficient hypertonic salt extraction procedure. These cell lines offer an abundant and genetically uniform source which permits the isolation of HL-A antigens in sufficiently large amounts to facilitate their rigorous biological and chemical characterization. The solubilized antigens are immunologically potent and specifically block the reactions of cytotoxic alloantisera in a pattern identical to that found upon typing of the cell donor's peripheral lymphocytes. The antigenic extracts can be most effectively fractionated by preparative polyacrylamide electrophoresis in discontinous buffer systems. The electrophoretically homogeneous antigenic component is a protein without lipid or carbohydrate moiety detectable at the 1% level. The purified HL-A antigen consists of a single polypeptide chain of molecular weight 31,000 containing two intrachain disulfide bridges. There are reproducible amino acid and peptide differences among antigens isolated from cultured lymphoid cells derived from histoincompatible individuals, suggesting that genes at certain histocompatibility loci directly code the synthesis of polypeptide chains. Further information on the chemical structure of HL-A antigens will clarify the nature of the determinants of histocompatibility and provide a better understanding of the mechanism of allograft rejection and ultimately of cell surface structure and function.

ACKNOWLEDGMENT

This is publication No. 536 from the Department of Experimental Pathology, Scripps Clinic and Research Foundation, La Jolla, California. This work was supported by United States Public Health Service Grants AI 10180 and CA 10596 from the National Institutes of Health and Grant 70-615 from the American Heart Association, Inc.

REFERENCES

Batchelor, R., and Selwood, N. (1970). In Terasaki, P. (ed.), *Histocompatibility Testing,* Munksgaard, Copenhagen, p. 243.
Boyse, E. A., Old, L. J., and Stockert, E. (1968). *Proc. Natl. Acad. Sci.* **60**:886.
Burnet, M. (1970). *Immunological Surveillance,* Sydney.
Burton, K. (1956). *Biochem. J.* **62**:315.
Ceppellini, R., Curtoni, E. S., Mattiuz, P. L., Miggaino, V., Scudeller, G., and Serra, A. (1968). In Curtoni, E. S., Mattiuz, P. L., and Tosi, R. M. (eds.), *Histocompatibility Testing,* Munksgaard, Copenhagen, pp. 149-185.
Cerottini, J. C., and Brunner, K. T. (1967). *Immunology* **13**:395.

Dandliker, W. B. R., Alonso, V. A., de Saussure, F., Kierszenbaum, F., Levison, S. A., and Shapiro, H. C. (1967). *Biochemistry* 6:1460.
Dausset, J., Walford, R. L., Colombani, J., Legrand, L., Feingold, N., Barge, A., and Rapaport, F. T. (1969). *Transplant. Proc.* 1:331.
Davies, D. A. L. (1967). *Transplantation* 5:31.
Davies, D. A. L. (1968). In Dausset, J., Hamburger, J., and Mathe, G. (eds.), *Advances in Transplantation,* Munksgaard, Copenhagen, pp. 275-279.
Davis, W. C., and Silverman, L. (1968). *Transplantation* 6:535.
Gervais, A. G. (1968). *Transplantation* 6:261.
Hatefi, Y., and Hanstein, W. G. (1969). *Proc. Natl. Acad. Sci.* 62:1129.
Hatefi, Y., and Hanstein, W. G. (1970). *Arch. Biochem. Biophys.* 138:73.
Haughton, G. (1964). *Transplantation* 2:251.
Hunter, M. S., and Commerford, S. L. (1961). *Biochim. Biophys. Acta* 47:580.
Jerne, N. (1971). *Europ. J. Immunol.* 1:1.
Kahan, B. D. (1965). *Proc. Natl. Acad. Sci.* 53:153.
Kahan, B. D., and Reisfeld, R. A. (1967). *Proc. Natl. Acad. Sci.* 58:1430.
Kahan, B. D., and Reisfeld, R. A. (1968a). *J. Immunol.* 101:237.
Kahan, B. D., and Reisfeld, R. A. (1968b). In Dausset, J., Hamburger, J., and Mathe, G. (eds.), *Advances in Transplantation,* Munksgaard, Copenhagen, p. 295.
Kahan, B. D., and Reisfeld, R. A. (1969a). *Science* 164:514.
Kahan, B. D., and Reisfeld, R. A. (1969b). *Proc. Soc. Exptl. Biol. Med.* 130:765.
Kahan, B. D., and Reisfeld, R. A. (1969c). *Transplant. Proc.* 1:483.
Kahan, B. D., and Reisfeld, R. A. (1971). *Bacteriol. Rev.* 35:59.
Kahan, B. D., Reisfeld, R. A., Epstein, L. B., and Southworth, J. G. (1967). In Curtoni, E. S., Mattiuz, P. L., and Tosi, R. M. (eds.), *Histocompatibility Testing,* Munksgaard, Copenhagen, pp. 295-299.
Kahan, B. D., Reisfeld, R. A., Pellegrino, M., Curtoni, E. S., Mattiuz, P. L., and Ceppellini, R. (1968). *Proc. Natl. Acad. Sci.* 61:897.
Kahan, B. D., Pellegrino, M., Papermaster, B. W., and Reisfeld, R. A. (1971). *Transplant. Proc.* 3:227.
Kandutsch, A. A. (1960). *Cancer Res.* 20:262.
Kandutsch, A. A. (1963). *Transplantation* 1:201.
Klotz, I. M., and Farnham, S. B. (1968). *Biochemistry* 7:3879.
Kourilsky, F. M., Silvestre, D., Levy, J. P., Dausset, J., Nicolai, M. G., and Senik, A. (1971). *J. Immunol.* 106:454.
Mann, D. L., and Levy, R. (1971). *Fed. Proc.* 30:2767.
Mann, D. L., Rogentine, G. N., and Fahey, J. L. (1968). *Nature (Lond.)* 217:1180.
Mann, D. L., Rogentine, G. N., Fahey, J. L., and Nathenson, S. G. (1969). *J. Immunol.* 103:282.
Mason, S., and Warner, N. L. (1970). *J. Immunol.* 104:762.
Medawar, P. B. (1963). *Transplantation* 1:21.
Möller, G. (1961). *J. Exptl. Med.* 114:415.
Moore, G. E., Gerner, R. F., and Franklin, H. A. (1967). *J. A. M. A.* 199:519.
Muramatsu, T., and Nathenson, S. G. (1971). *Fed. Proc.* 30:2768.
Nathenson, S. G., and Davies, D. A. L. (1966). *Proc. Natl. Acad. Sci.* 56:476.
Ornstein, L. (1964). *Ann. N.Y. Acad. Sci.* 121:321.
Ozer, J. H., and Hoelzl-Wallach, D. F. (1967). *Transplantation* 5:652.
Raymond, S., and Weintraub, L. (1959). *Science* 130:711.
Reif, A. (1966). *Immunochemistry* 3:267.
Reisfeld, R. A., and Kahan, B. D. (1970a). In Dixon, F. J., and Kunkel, H. (eds.), *Advances in Immunology,* Vol. 12, pp. 117-191.
Reisfeld, R. A., and Kahan, B. D. (1970b). *Fed. Proc.* 29:2034.
Reisfeld, R. A., and Kahan, B. D. (1971). *Transplant. Rev.* (in press).
Reisfeld, R. A., Pellegrino, M., Papermaster, B. W., and Kahan, B. D. (1970a). J. Immunol. 104:560.

Reisfeld, R. A., Pellegrino, M., Papermaster, B. W., and Kahan, B. D. (1970*b*). In Terasaki, P. I. (ed.), *Histocompatibility Testing,* Munksgaard, Copenhagen.

Reisfeld, R. A., Pellegrino, M., and Kahan, B. D. (1971*a*). *Science* 172:1134.

Reisfeld, R. A., Pellegrino, M., Papermaster, B. W., and Kahan, B. D. (1971*b*). *Proc. VI Internat. Immunopathol. Symp.* (in press).

Rodbard, D., and Chrambach, A. (1971). *Anal. Biochem.* 40:95.

Sanderson, A. R. (1968). *Nature (Lond.)* 220:192.

Sanderson, A. R., Cresswell, P., and Welsh, K. I. (1971). *Nature (Lond.)* 230:8.

Shimada, A., and Nathenson, S. G. (1969). *Biochemistry* 8:4048.

Terasaki, P. I., and McClelland, J. D. (1964). *Nature (Lond.)* 204:998.

Thomas, L. (1959). In *Cellular and Humoral Aspects of the Hypersensitive State,* Wiley and Sons, London, pp. 451-468.

Wallach, D. F. H., and Gordon, A. S. (1969). In Smith, R. T. and Good, R. A. (eds.), *Cellular Recognition,* Appleton, New York, p. 3.

Bacterial Flagellin as an Antigen and Immunogen

Christopher R. Parish and Gordon L. Ada

Department of Microbiology
John Curtin School of Medical Research
The Australian National University
Canberra, Australia

I. INTRODUCTION

The ability to recognize "foreignness" seems to have been a very early development in phylogeny. It has been shown that discrimination can occur even with single-cell organisms. It is only in vertebrates, however, that the ability has been highly developed as regards both specificity and variety in the level of discrimination. This early form of discrimination was at the level of ingestion and destruction—a process which persists and is of prime importance even in vertebrates. To this was added the ability of multicell organisms to recognize "foreign" cells, with minor differences in specificity between self and nonself being of major importance. On this basis, the vertebrates again developed a rather more complex system—the ability to form a particular class of substance having great specificity, i.e., antibodies, against the inducing substances. In mammals, this culminated in the development of a variety of classes of antibodies, each one of which has specific properties which are of relevance in particular circumstances.

All these "immune" responses depend for their initiation on the inducing substance—the antigen. Most investigations have been carried out in mammals, and the very complexity of the system here makes it all the more difficult to study the inductive role of the antigen. Though the study of all these different facets has been facilitated by using a variety of inducing substances such as proteins substituted with haptens and synthetic polypeptides, it would help our understanding of the interrelationships of the different responses if a single parent substance and known derivatives of this were used.

When some of the investigations to be described in this chapter were begun

77

(Ada *et al.*, 1964), flagellin from *Salmonella* organisms was chosen as a suitable protein antigen because it was known to be highly immunogenic in rodents, even when injected in the absence of adjuvants. This property has proved to be of paramount importance, for a number of reasons. Initially, it enabled an investigation into the *in vivo* fate of the antigen under conditions which, it could be argued, were not too dissimilar from "physiological." Secondly, because it is highly immunogenic, the effect of chemical or physical modification on the biological properties could easily be followed, usually as a decrease in immunogenicity but frequently with an enhancement of another property, such as tolerogenicity. Thirdly, it is, to date, the only well-defined protein antigen which has been used with great success in tissue culture investigations of the immune response.

For brevity, we will restrict our discussion to the flagellin of *Salmonella* organisms. Furthermore, as some of the work has recently been summarized elsewhere (Nossal and Ada, 1971), only the more recent observations will be given in detail.

II. PROPERTIES OF FLAGELLIN FROM *SALMONELLA* ORGANISMS

A. Physical and Chemical Properties

Bacterial flagellin is a protein with a molecular weight of about 40,000 (Joys, 1968; Parish and Marchalonis, 1970) and constitutes the subunit of the flagellum, the organelle of locomotion in *Salmonella* organisms. Viewed under the electron microscope, flagella appear as long, thin filaments, attached at one end to the bacterial body, this area of attachment frequently forming a hook (Abram *et al.*, 1966). The bacterial flagellum is composed of flagellin molecules apparently arranged in a number (probably eight to ten) of helical subfibrils (Lowy and Hanson, 1965). Flagella also contain some nonproteinaceous material which is composed predominantly of carbohydrate (e.g., Ada *et al.*, 1964). Flagella are readily dissociated into their flagellin subunits by treatment with heat, extremes of pH, or concentrated urea solutions (Abram and Koffler, 1964; Martinez *et al.*, 1967). During acid dissociation of flagella, the bulk of the carbohydrate remains insoluble, yielding a solution of almost pure flagellin. Flagellin readily repolymerizes in the presence of moderately high salt concentrations (Ada *et al.*, 1964) or flagellin "seed" (Asakura *et al.*, 1964) to form a structure very similar to the flagellum, but this repolymerized flagellin lacks the nonprotein components. Thus the flagellin molecule can exist in three physical states—as dissociated flagellin ("monomer"), as a subunit incorporated in the flagella particle, or as repolymerized flagellin ("polymer").

The *Salmonella* flagellins are also characterized by their amino acid composition. There is an abundance of aspartic acid and alanine but no tryptophan, hydroxyproline, or cysteine, sometimes no histidine, and only small amounts of tyrosine, methionine, and proline (McDonough, 1965; Parish and Ada, 1969a). In addition, *Salmonella* flagellins are characterized by the presence of the unusual amino acid ε-*N*-methyllysine, which occurs in several but not all serotypes (Ambler and Rees, 1959).

B. Antigenic Properties

Flagella constitute the H antigenic specificity of *Salmonella*, this specificity being carried by the flagellin subunits (Iino and Lederberg, 1964). The H antigens of *Salmonella* flagella are complicated by the existence of phase variation, although monophasic strains are used for most immunological studies (e.g., *S. adelaide*). The genetic basis of phase variation is discussed in detail in two reviews (Iino and Lederberg, 1964; Joys, 1968).

It has been reported that the flagellin subunits possess antigenic determinants which are not present on flagella (e.g., Joys, 1968). Flagellin from *S. adelaide* contains these flagellin-specific determinants, but they constitute a minor antigenic component (Pye, 1968). It also appears that flagella-specific determinants exist in some *Salmonella* strains and in some other bacterial species (Joys, 1968; Ichiki and Martinez, 1969), although there is no evidence that this type of determinant occurs in *S. adelaide* flagella.

In this laboratory, two main approaches have been used to examine the antigenic structure of *S. adelaide* flagellin: (1) the isolation and characterization of antigenically active fragments from chemical and enzymic digests of flagellin and (2) the chemical modification of side groups of flagellin and evaluation of the effect this has on the antigenic behavior of the protein.

III. DEGRADATION AND MODIFICATION OF FLAGELLIN

A. Antigenic Properties of Fragments Released from Flagellin

Antigenic fragments are recovered from *S. adelaide* flagellin following breakdown of the molecule by either trypsin, pepsin, or cyanogen bromide. Cyanogen bromide (CNBr) digestion leaves almost intact the antigenic determinants of the flagellin molecule (Parish *et al.*, 1969). Cyanogen bromide specifically and almost quantitatively splits the peptide chain at the methionine residues, and, as *S. adelaide* flagellin contains three methionines per mole, four CNBr fragments are obtained. These fragments were arbitrarily called A, B, C, and D in order of decreasing size. They have molecular weights ranging from 18,000 to 4500, and their sequence in the flagellin molecule is BADC. Several serological tests revealed that fragment A carries all of the antigenic determinants of flagellin (Parish et al., 1969), although the affinity of these determinants for antiflagellin antibodies is slightly reduced (Parish, 1971a). Fragments B, C, and D are antigenically inert, and one or more of these may be concerned in the polymerization of the monomer to form the polymer.

When flagellin is digested with pepsin at a suboptimal pH (pH 5.0, 2.5 hr) or with trypsin at a suboptimal temperature (25°C), large, antigenically active fragments are released (Ichiki and Parish, submitted for publication). Peptic digestion produces three or four antigenic fragments with molecular weights between 13,000 and 15,000, whereas trypsin releases two antigenic peptides with molecular weights approximating 16,000. Prolonged peptic digestion (pH 5.0, 24 hr) completely degrades flagellin, and, similarly, tryptic digestion at a higher temperature (37°C) results in extensive breakdown. Most of the antigenic determinants of flagellin are localized on the large tryptic and peptic peptides, although sensitive serological tests indicated the loss or a weakening of the activity of one or more determinants. Additional degradation studies revealed that the antigenic peptic peptides are derived solely from within fragment A, but this is not the case with the tryptic peptides, which overlap between fragment A and either fragment B or fragment D. From comparison of the amino acid compositions of the antigenically active tryptic and peptic peptides and the CNBr fragments, it was concluded that neither the methionine nor the arginine residues of flagellin play an important antigenic role.

B. Antigenic Properties of Chemically Modified Flagellin

The antigenicity of S. adelaide flagellin has been modified by a range of substitution procedures, such as dinitrophenylation, maleylation, and acetoacetylation. To date, the acetoacetyl derivatives of flagellin have been the ones most carefully characterized (Parish, 1971a). A range of flagellin preparations has been obtained in which flagellin has been substituted to varying extents with acetoacetyl groups. The antigenic activity of these derivatives was quantitated, and it was found that the lysine residues of flagellin are not intimately associated with the antigenic determinants of the molecule, whereas the readily substituted hydroxyl groups (presumed to be tyrosine residues) are antigenically very important.

There is some evidence that iodination of the tyrosine residues may also modify antigenicity. On the other hand, oxidation of the methionine residues by chloramine-T does not affect antigenic activity even though preparations of flagellin so treated fail to polymerize (Parish, 1969).

From these chemical and antigenic studies, it was concluded that the flagellin molecule possesses a structural polarity, part of the molecule being antigenically important and another part being associated with the polymerizing properties of the molecule.

IV. *IN VIVO* IMMUNOGENICITY OF FLAGELLIN AND ITS DERIVATIVES

A. Immunogenicity of Flagellin and Polymerized Flagellin

One of the major reasons for choosing flagellin as an antigen is its high immunogenicity. It has been found that the injection into rats of as little as 10 ng of *S. adelaide* flagellin in saline causes the production of detectable amounts of antiflagellin antibody (Nossal *et al.*, 1964). Higher antibody responses are observed when flagellin in adjuvants is injected (Lind, 1968). The high immunogenicity of the flagellar antigens is probably not due to the pre-exposure of animals to these antigens, as germ-free rats respond just as well as conventional rats (Miller *et al.*, 1968). In adult rats, a single injection of polymer induces both an IgM and an IgG response, whereas flagellin induces predominantly an IgG response. It is of interest that doses of polymer or flagellin which induce high titers of hemagglutinating and bacterial immobilizing antibody fail to induce detectable Arthus reactions (Parish, 1971*b*).

In contrast, flagellin induces a comparatively weak delayed-type hypersensitivity response. The response seen is induced equally well by injecting flagellin in saline or in adjuvant (Parish, 1971*b*). It seems that the injection of flagellin and polymer favors the induction of antibody formation rather than a cell-mediated immune response.

Neither polymer nor flagellin can induce antibody tolerance in adult rats, even when multiple doses of antigen are administered. However, it has been shown that antibody tolerance to flagellin can be induced in adult rats and mice by multiple injections of antigen during the period of recovery either from prolonged thoracic duct cannulation or from treatment with antilymphocyte serum (Shellam, 1969*b*). Furthermore, both flagellin and polymer can induce antibody tolerance in neonatal rats, although tolerance only occurs following multiple injections of antigen at two widely spaced dose levels. Remarkably low concentrations of antigen induce low-dose tolerance, about 10^{-14} M in the case of flagellin and 10^{-20} M for polymer (Shellam and Nossal, 1968; Shellam, 1969*a,b*).

There is some evidence that the flagellar antigens are thymus dependent in both rats (Steward, 1969; Lind, 1970) and mice (Miller, 1962; Humphrey *et al.*, 1964; Davies *et al.*, 1970), although conflicting results have been obtained from some neonatal thymectomy studies (Pinnas and Fitch, 1966). It appears that the strain of the experimental animal, the antigenic specificity of the flagellin, and the physical state of the flagellar antigens (i.e., monomer or polymer) determine whether or not thymus dependence can be demonstrated.

Table I. Relationship Between Antigen Dose and Ability of Cyanogen
Bromide-Digested Flagellin to Induce Antibody Formation and
Delayed-Type Hypersensitivity in Adult Rats[a]

| | | Subsequent immune response to flagellin (100 μg in saline) | |
| | | Antibody response (28 days) | Delayed-type hypersensitivity |
Antigen	Dose		
Nil	–	180	2.1
CNBr digest	100 μg/day	70[b]	3.4[c]
	100 ng/day	960[c]	0.5[b]
	1 pg/day	80[b]	4.0[c]

[a] Adult rats were injected for 28 days with the different doses of the CNBr digest and then
challenged with 100 μg of flagellin in saline. Four weeks later, delayed-type hypersensitivity was elicited with 0.5 μg of flagellin in saline. Antibody titers are expressed as
reciprocal of dilution. Hypersensitivity measurements represent 24 hr footpad swellings
(1/10 mm).
[b] Significant tolerance compared with control animals.
[c] Significant enhancement compared with control animals.

B. Immunogenicity of Fragmented Flagellin

Fragmentation of flagellin by cyanogen bromide or pepsin greatly modifies
the immunological properties of the molecule. Cyanogen bromide digestion
significantly reduced the antibody-inducing capacity of flagellin (Parish and Ada,
1969b) but substantially enhanced the delayed-type hypersensitivity induced
(Parish, 1971b). It was subsequently found that fragment A, previously shown
to be the antigenically active fragment in a CNBr digest, was also the ingredient
of the digest active in *in vivo* experiments (see Table II). Multiple injections of
the CNBr digest at two widely spaced dose levels (100 μg/day and 0.1 pg/day)
induced antibody tolerance in adult rats, whereas intermediate doses resulted in
enhanced antibody responses (Ada and Parish, 1968). Further studies of this
system have revealed an inverse relationship between the resulting antibody
response and the level of delayed-type hypersensitivity (Table I). High- and
low-dose antibody tolerance is accompanied by enhanced cell-mediated immunity, whereas an enhanced antibody response reflects cellular immunity
tolerance. In contrast, neonatal injections of the CNBr digest induced tolerance
at the level of *both* humoral and cell-mediated immunity (Parish and Liew,
submitted for publication).

Peptic digestion of flagellin also destroyed the ability of flagellin to induce
antibody formation. However, this degradation not only did not destroy but

rather tended to enhance the ability to sensitize rats to give a delayed-type hypersensitivity reaction to flagellin (Table II) (Ichiki and Parish, submitted for publication). In fact, it was found that the most highly degraded preparations induced the best cell-mediated immunity, even though these preparations had little or no affinity for antiflagellin antibodies. Even so, the peptides which were capable of inducing delayed-type hypersensitivity were found to be derived from the region of flagellin which bound antiflagellin antibodies (i.e., fragment A and the pepsin-resistant peptides).

C. Immunogenicity of Chemically Modified Flagellin

Table III summarizes the immunological properties of some of the aceto-acetylated derivatives of flagellin. Increasing acetoacetylation steadily destroyed

Table II. Ability of the Peptic and Cyanogen Bromide Fragments of Flagellin to Induce Antibody Formation and Delayed-Type Hypersensitivity[a]

Priming antigen (1 μg) + FCA	Molecular weight	Relative antigenic activity[b] (K_{rel})	Delayed-type hyper-sensitivity	Antibody response (28 days)
Saline	–	–	0.9	< 2.5
Flagellin	40,000	1.0	2.2[c]	1280
Fragment A	18,000	5.7×10^{-1}	3.3[c]	240
Fragments B, C, D	4500–12,000	$>10^{-5}$	0.9	<2.5
High mol. wt. peptic peptides (2.5 hr)	13,000–15,000	2.9×10^{-1}	2.1[c]	2.5
Low mol. wt. peptic peptides (2.5 hr)	<4000	$>10^{-3}$	2.9[c]	< 2.5
Degraded high mol. wt. peptic peptides	<4000	$>10^{-3}$	2.9[c]	< 2.5

[a] Adult rats were immunized and 28 days later injected with 0.5 μg of flagellin in saline and tested for delayed-type hypersensitivity. Antibody titers are expressed as reciprocal of dilution. Hypersensitivity measurements represent 24 hr footpad swellings (1/10 mm).

[b] The antigenic activity of the antigens compared with unmodified flagellin.

[c] Significant delayed-type hypersensitivity compared with saline control.

Table III. Relationship Between Antigenic Activity and Ability
of Acetoacetylated Flagellin to Induce Antibody Formation
and Delayed-Type Hypersensitivity[a]

Antigen initially injected (1 μg)	Relative antigenic activity[b] (K_{rel})	Delayed-type hyper-sensitivity	Primary antibody response (day 35)	Antibody tolerance (day 28)
Saline	–	0.5	< 2.5	608
Flagellin	1.0	2.3	9730	11,260
Flagellin (8.0 acetoacetyl[c] groups)	6.1×10^{-1}	4.5^d	1220	4,860
Flagellin (10.8 acetoacetyl groups	3.45×10^{-1}	6.0^d	5	832
Flagellin (16.8 acetoacetyl groups)	6.8×10^{-3}	7.8^d	< 2.5	76^e
Flagellin (20.6 acetoacetyl groups)	6.8×10^{-10}	4.1^d	< 2.5	544

[a] Adult rats were injected with 1 μg of each antigen in Freund's complete adjuvant. Animals were challenged 35 days later with 1 μg of flagellin in saline (hind footpads) and tested for delayed-type hypersensitivity and antibody tolerance. Antibody titers are expressed as reciprocal of dilution. Hypersensitivity measurements represent 24 footpad swellings (1/10 mm). For further details, see Parish (1971b,c).

[b] The antigenic activity of the antigens compared with unmodified flagellin (Parish, 1971a).

[c] Number of acetoacetyl groups attached per mole.

[d] Significantly higher delayed-type hypersensitivity than flagellin control.

[e] Significant antibody tolerance compared with control (saline primed).

the ability of flagellin to initiate antibody formation but *enhanced* the ability of the molecule both to induce antibody tolerance and to sensitize for a flagellin-specific cell-mediated immunity (Parish, 1971a,b,c). In fact, it appears that in adult animals and for this antigen, antibody formation and cell-mediated immunity may well be opposing immunological processes.

The explanation proposed for these effects is that the relative affinity of the antigen for the cell receptor (as estimated *in vitro* by measurement with antibody) determines the type of immune response which will predominate. A decrease in relative affinity due to substitution with acetoacetyl groups favored the induction of a cell-mediated immune response. The immunological behavior of fragment A and of the pepsin-resistant peptides was consistent in this way with their antigenic activities (Table II), although the peptic peptides produced a lower cell-mediated immune response than would be expected on this basis, possibly because some antigenic determinants were completely destroyed by digestion. A finding consistent with this notion was that mild oxidation of flagellin with chloramine-T, which failed to affect the antigenicity of the molecule, did not alter its immunogenicity (Parish, 1969).

In contrast to the above studies with adult rats, tolerance in neonatal rats was induced by acetoacetylated flagellin (16.8 acetoacetyl groups per mole) at the level of both humoral and cell-mediated immunity (Parish, 1971*b*).

V. *IN VIVO* LOCALIZATION PATTERNS OF FLAGELLIN AND ITS DERIVATIVES

The finding that the flagellar proteins were immunogenic at very low doses was made at about the same time as both "carrier-free" preparations of iodine-125 and methods for the direct iodination of protein became available. Using these low concentrations of antigen, the following points were established (Nossal and Ada, 1971): (1) Of the antigen (monomer or polymer) present in a draining lymph node at any time after subcutaneous injection, a large proportion was in macrophages, where it persisted. (2) Depending upon the immune status of the animal, a varying proportion of the antigen was present at the surface of dendritic cells in lymphoid follicles. The mechanism of the binding of the antigen to these cells was later shown to be antibody mediated. Follicular localization of antigen appears to be a very neat method for the extracellular concentration of antigen; as such, it was invoked as a possible explanation for the phenomenon of ultra low zone tolerance (see Table I), where it seemed to be mandatory to postulate a mechanism for concentrating antigen (Ada and Parish, 1968). However, there is no decisive evidence which pinpoints the role in immune processes which antigen in either of these areas might play; similarly, the possibility has not been eliminated that antigen associated with either of these cell types might be involved in more than one type of immune response. The difficulties of interpretation in this area are very great.

Perhaps most interesting was the failure to find the injected antigen associated with lymphocytes in the lymphoid organs of unprimed animals. This seemed to favor the interpretation that the lymphocyte received the "antigenic message" via macrophages or dendritic cells. Such an interpretation was certainly consistent with other work with isolated macrophages, suggesting an important role of this nature for this cell. It now seems that in the early work the search for labeled lymphocytes may not have been sufficiently precise, because very recently, by use of electron microscopic radioautography, injected labeled polymerized flagellin has been seen to react directly with lymphocytes present in the marginal zone of rat spleen (Mitchell and Abbot, personal communication). However, the early work did establish very clearly that after injection of labeled antigen, cells actively secreting specific antibody contained little labeled antigen (estimated to be less than four molecules per cell; Nossal *et al.,* 1965). McDevitt *et al.* (1966) made similar observations using a radioiodinated synthetic polypeptide as antigen.

The localization patterns of the flagellin derivatives have not been examined in detail. In accord with its smaller size, fragment A was found to be trapped less efficiently in the draining lymph node than was flagellin, though the localization in the lymphoid follicles (due to "natural" antibody) appeared to be unaffected (Ada and Parish, 1968).

VI. *IN VITRO* BEHAVIOR OF FLAGELLIN AND ITS DERIVATIVES

A. Induction of Immune Responses

In 1966, Mishell and Dutton described a tissue culture technique in which antibody to red blood cells was produced by separated cells from spleens of unimmunized mice. A variant of this technique was developed by Marbrook (1967), and this latter procedure, slightly modified, has been used extensively by Diener and colleagues to study the *in vitro* induction of immune responses by flagellar antigens. Many of the results obtained to date are summarized in Table IV and are compared with *in vivo* results. It should be pointed out that the *in vitro* results were obtained using mouse (CBA strain) spleen cells, whereas most of the *in vivo* results quoted above were obtained in rats. It is unlikely that this difference in hosts affects the validity of the main conclusions drawn.

As seen in Table IV, there is a hierachy in the *in vivo* immunogenicity (antibody induction) of flagellar antigens, ranging from the highly immunogenic polymer to the inactive acetoacetylated flagellin. In contrast, of these substances, only the polymer is active *in vitro,* the monomer flagellin showing only marginal activity. Whereas a wide range of dose levels of the polymer is immunogenic *in vivo,* only a narrow range of dose levels induces antibody production *in vitro;* dose levels 100-fold or more higher cause specific antibody tolerance. Equally striking are other results obtained in the tolerance experiments. In adult rats, antibody tolerance, either high or low zone, is obtained only with the derivatives, either fragment A or acetoacetylated flagellin. In tissue culture, both of these derivatives are inactive *unless* specific antibody is added in appropriate proportions relative to the antigen concentration. It is of further interest that, as in *in vivo* experiments, two zones of antibody tolerance have been demonstrated in tissue culture using fragment A and antibody. It is to be

Table IV. *In Vitro* and *in Vivo* Responses to Flagellar Antigens

Immune response		Polymerized flagellin	Polymerized flagellin + antibody	Flagellin	Fragment A	Fragment A + antibody	Aceto-acetylated flagellin	Aceto-acetylated flagellin + antibody
Antibody production	*In vivo*	+++		++	±		−	
	In vitro[a]	+++		±	−		−	
Cell-mediated immunity	*In vivo*	±		+	++		+++	
	In vitro	+++						
High zone tolerance (adult)	*In vivo*	−		−	++		++	
	In vitro	+++	+++	−	−	++	−[b]	+[b]
Low zone tolerance (adult)	*In vivo*	−		−	++	++[c]	−[b]	
	In vitro	−		−	−		−[b]	

[a] References to most of these results are quoted elsewhere (Nossal and Ada, 1971).
[b] Feldman and Parish (unpublished).
[c] Diener and Feldman (1970).

noted from Table IV that techniques remain to be developed for the *in vitro* demonstration of cell-mediated immune responses to these antigens.

In the preceding paragraph, the differences between the *in vivo* and the *in vitro* results have been stressed. It is equally important to stress that with the exception, to date, of the cell-mediated immune responses, essentially all the immune responses found *in vivo* have been reproduced in tissue culture, and this must be regarded as a remarkable achievement. The reasons for the different results become more obvious when we consider further work by Diener and colleagues.

We have noted above that *in vivo* the cells which demonstrably bind or react with injected labeled antigen are macrophages, dendritic cells, and lymphocytes (presumably, these may be bone marrow derived and thymus derived). When spleen cell suspensions are made, it seems that dendritic cells are poorly, if at all, recovered, so that whatever role antigen on dendritic cells may play *in vivo,* this is not reproduced *in vitro.* As antigen is bound to dendritic cells by antibody, this may explain the need to add antibody *in vitro* in order to obtain tolerance when fragment A is used as the antigen. On the other hand, spleen suspensions do contain macrophages. Do these cells play a role in the induction of antibody formation to the polymer? This question has been answered by Shortman and colleagues (Shortman *et al.,* 1970; Diener *et al.,* 1970). Cells were fractionated either by equilibrium centrifugation or by passage through glass bead columns. By either procedure, it was found possible to obtain spleen cell suspensions greatly depleted with respect to phagocytic cells. The depleted populations were found both to support adequately antibody production to the polymer and to allow the induction of antibody tolerance. Thus it seemed that, for this antigen, reaction with macrophages was unnecessary to obtain either of these responses. It follows that if the antigen which reacted with macrophages *in vivo* does play a role in the induction of either antibody production or tolerance, such a role would be mechanical in nature rather than one involving information transfer by nonantigenic material from macrophage to lymphocyte.

B. Reaction Between Antigen and Lymphocytes

These results support the belief which has been current now for some time that the critical event in the immune response is the reaction between antigen and lymphocyte—an event which determines whether either antibody production and secretion, cell-mediated immunity, or tolerance will ensue. In view of this, it is surprising that perhaps the most direct way of studying this reaction was not described until 1967. At that time, Naor and Sulitzeanu reported experiments in which suspensions of mouse spleen cells were allowed to react at 4°C with low concentrations of bovine serum albumin, trace labeled with iodine-125. The prediction of Burnet's clonal selection theory was confirmed;

namely, a small proportion only of lymphocytes was found to bind the antigen, and within this subpopulation there was a hierachy, with some cells binding more antigen than others. This reaction has been investigated in considerable detail by others (Byrt and Ada, 1969; Ada *et al.*, 1970; Humphrey and Keller, 1970) and has been recently reviewed (Ada, 1970). Some findings which are of relevance here are the following: (1) Antigen-binding lymphocytes for a particular antigen occur at frequencies ranging from about 10^{-2} to at least 10^{-5}. (2) Binding of labeled antigen to mouse spleen cells can be inhibited if the cells are first exposed to antimouse immunoglobulin, supporting the concept that the receptor for antigen is immunoglobulin. In most spleen cells, the predominant immunoglobulin is IgM. (3) Binding of the antigen to the lymphocyte can be inhibited by excess unlabeled antigen of the same specificity. (4) For at least a proportion of the binding cells, the reaction with antigen is specific. This was demonstrated by the "suicide experiment" (Ada and Byrt, 1969) in which polymerized flagellin was added to mouse spleen cells, the binding of antigen allowed to take place, and then the unbound antigen removed. Binding of a labeled specific antigen to a population of cells was found to abrogate selectively the ability of the cell population to produce antibody to that antigen.

The binding of flagellin and its derivatives to unimmunized rat spleen cells has been studied (Ada and Cooper, 1971). Under comparable conditions, fragment A bound to fewer cells than did flagellin, as expected. Contrary to expectation, acetoacetylated flagellin bound to many more spleen cells than did flagellin, and appropriate tests suggested that this reaction was specific. The reason for the enhanced reactivity of acetoacetylated flagellin with spleen cells and to what extent it is relevant to the biological activity of the protein remain to be determined.

VII. ANTIGEN AS A REGULATOR OF CELL BEHAVIOR

Antigen is probably the most important regulator of the behavior of lymphocytes, determining whether these are made tolerant (deletion or inactivation?) or proliferate and differentiate. Recent work shows that the behavior of other cell types may also be affected by antigen. Of the antigens tested, polymerized flagellin appears to be very effective. This antigen markedly stimulates the proliferation of granulocyte and macrophage precursors, both *in vivo* and *in vitro*. This stimulation may be due to the production of a colony-stimulating factor or another promoting factor (McNeill, 1970*a,b;* Metcalf, 1971).

VIII. DISCUSSION AND CONCLUSIONS

It has become very clear that the different immune responses which can be induced by antigens are closely interrelated and that one or two of them cannot be adequately studied in isolation. Bearing this in mind, in this article we have tried to point out how extraordinarily useful the flagellar antigens have been in elucidating the immune system. They have been used to study all major aspects of the immune response *in vivo* and antibody production and tolerance *in vitro*. Use of these antigens has led to the demonstration of many immunological phenomena and suggested a number of important concepts. Some of them, and their implications, are worth emphasizing.

The finding that antibody-secreting cells contained little, if any antigen, was one of the more substantial nails in the coffin of the template theory of antibody formation. It is entirely in line with all the recent work indicating that the interaction of antigen with the lymphocyte surface receptor is probably the trigger which sets in motion the cell's response. The specificity of the immune response is determined by this interaction, as shown by the suicide experiment. The demonstration of ultra low zone tolerance both in neonatal and adult rats is, we believe, important because it places great restrictions on the type of mechanism which can be invoked to explain the phenomenon and suggests that mechanisms other than "paralysis" are responsible. The conundrum is: How can such tiny amounts of antigen—tiny even in molecular terms—induce a state of tolerance, particularly as probably less than 1% is trapped in lymphoid organs?

One of the more recent findings is one of the more important—the demonstration of an inverse relationship between antibody production and cell-mediated immunity. To what extent this may be a general finding remains to be determined, but it does raise a most important practical aspect of immunology—the possibility of immunological engineering. For example, modification of a vaccine might be used to cause an immunity which is predominantly humoral *or* cell-mediated in nature.

One of the important features of the flagellar antigenic system was its availability in a monomeric or polymeric form and more recently as fragments of flagellin. Initially, there seemed to be a number of situations in which the "size" of the antigen administered was a decisive factor in determining the type of immune response. For example, injection of polymer gave both an IgM and an IgG response, whereas the response to monomer was IgG alone.

Similarly, antigenic fragments of flagellin were poorer inducers of antibody formation than undegraded flagellin and, in fact, were effective inducers of antibody tolerance in adult rats. On the other hand, decreasing molecular size tended to enhance the cellular immune responses induced; thus the order of effectiveness was degraded flagellin, monomer, polymer. Molecular size was also an important factor in determining the *in vitro* immune response, polymerized flagellin being the only form of flagellin which could directly induce either an antibody response or antibody tolerance. It is now realized to be unlikely that these different effects can be attributed to size alone. Additional factors are involved. One of these is the affinity of antigen for cell receptors, high-affinity antigen favoring antibody formation and possibly cellular immunity tolerance and low-affinity antigen favoring antibody tolerance and enhanced cellular immunity. We also need to superimpose an additional effect to explain the different response engendered by polymer compared to monomer. It seems possible that the binding of antigenic determinants (e.g., polymer) to adjacent receptors on a cell surface may more effectively "trigger" an immunocompetent cell than equal numbers of unconnected binding events (e.g., monomer). The demonstration that flagellin and fragment A, which are immunologically inert *in vitro,* could induce antibody tolerance in the presence of specific antibody suggested that the antibody may act by causing a local concentration of antigenic determinants (Ada, 1970). Furthermore, repeating antigenic determinants would be expected to favor cell-cell interaction and therefore antibody production.

One further comment is worthwhile, and this concerns the role of the macrophage in the immune response. The *in vitro* experiments of Shortman and colleagues suggest that the main role of the macrophage is a mechanical breakdown of the antigen. To this should be added a possible role of effective presentation of antigen to lymphocytes (e.g., see Nossal and Ada, 1971, for review). It now seems very likely that one other function that has not been sufficiently realized previously should be added. This is that extensive breakdown of antigen to small fragments, possibly monodeterminant or with a reduced number of determinants, will favor a cell-mediated immune response (see also Pearson and Raffel, 1971). It may also help to explain why both polymer and monomer are poor inducers of cell-mediated immune responses. The antigenic determinants of polymer and monomer are present in regions of the molecule which are relatively resistant to proteases, and it was noticed that both antigens persist for long periods in lymph node macrophages.

In conclusion, we have attempted to outline major results obtained using the flagellar antigens and their derivatives. We wish to stress two points: (1) In future work, studies on the immunogenic properties of antigens should consider all parameters of the immune response. (2) We believe that the flagellar antigens offer a system which already has led to a clarification of some of the inter-relationships among the different immune systems and one which should, in the future, be increasingly useful.

REFERENCES

Abram, D., and Koffler, H. (1964). *J. Mol. Biol.* 9:168.
Abram, D., Vatter, A. E., and Koffler, H. (1966). *J. Bacteriol.* 91:2045.
Ada, G. L. (1970). *Transplant. Rev.* 5:105.
Ada, G. L., and Byrt, P. (1969). *Nature (Lond.)* 222:1291.
Ada, G. L., and Cooper, M. G. (1971). *Ann. N.Y. Acad. Sci.* 181:96.
Ada, G. L., and Parish, C. R. (1968). *Proc. Natl. Acad. Sci.* 61:556.

92   Parish and Ada

Ada, G. L., Nossal, G. J. V., Pye, J., and Abbot, A. (1964). *Aust. J. Exptl. Biol. Med. Sci.* 42:267.
Ada, G. L., Byrt, P., Mandel, T., and Warner, N. L. (1970). In Sterzl, J., and Riha, M. (eds.), *Developmental Aspects of Antibody Formation and Structure,* Academic Press, New York.
Ambler, R. P., and Rees, M. W. (1959). *Nature (Lond.)* 184:56.
Asakura, S., Eguchi, G., and Iino, T. (1964). *J. Mol. Biol.* 10:42.
Byrt, P., and Ada, G. L. (1969). *Immunology* 17:503.
Davies, A. J. S., Carter, R. L., Leuchars, E., Wallis, V., and Dietrich, F. M. (1970). *Immunology* 19:945.
Diener, E., and Feldman, M. (1970). *J. Exptl. Med.* 132:31.
Diener, E., Shortman, K., and Russell, P. (1970). *Nature (Lond.)* 225:731.
Humphrey, J. H., and Keller, H. U. (1970). In Sterzl, J., and Riha, M. (eds.), *Developmental Aspects of Antibody Formation and Structure,* Academic Press, New York.
Humphrey, J. H., Parrott, D. M. V., and East, J. (1964). *Immunology* 7:419.
Ichiki, A. T., and Martinez, R. J. (1969). *J. Bacteriol.* 98:481.
Iino, T., and Lederberg, J. (1964). In *The World Problem of Salmonellosis,* Dr. W. Junk, The Hague, Netherlands.
Joys, T. M. (1968). *Antonie van Leeuwenhoek J. Microbiol. Serol.* 34:205.
Lind, P. E. (1968). *Aust. J. Exptl. Biol. Med. Sci.* 46:179.
Lind, P. E. (1970). *Internat. Arch. Allergy* 37:258.
Lowy, J., and Hanson, J. (1965). *J. Mol. Biol.* 11:293.
Marbrook, J. (1967). *Lancet* ii:1279.
Martinez, R. J., Brown, D. M., and Glazer, A. N. (1967). *J. Mol. Biol.* 28:45.
McDevitt, H. O., Askonas, B. A., Humphrey, J. H., Schechter, I., and Sela, M. (1966). *Immunology* 11:337.
McDonough, M. W. (1965). *J. Mol. Biol.* 13:342.
McNeill, T. A. (1970a). *Immunology* 18:39.
McNeill, T. A. (1970b). *Immunology* 18:61.
Metcalf, D. (1971). *Immunology* 21:427.
Miller, J. F. A. P. (1962). *Proc. Roy. Soc.* 156:415.
Miller, J. J., III, Johnsen, D. O., and Ada, G. L. (1968). *Nature (Lond.)* 217:1059.
Mishell, R. I., and Dutton, R. W. (1966). *Science* 153:1004.
Naor, D., and Sulitzeanu, D. (1967). *Nature (Lond.)* 214:687.
Nossal, G. J. V., and Ada, G. L. (1971). In Dixon, F. J., Jr., and Kunbel, H. G. (eds.), *Antigens, Lymphoid Cells and the Immune Response,* Academic Press, New York and London.
Nossal, G. J. V., Ada, G. L., and Austin, C. M. (1964). *Aust. J. Exptl. Biol. Med. Sci.* 42:283.
Nossal, G. J. V., Ada, G. L., and Austin, C. M. (1965). *J. Exptl. Med.* 121:945.
Parish, C. R. (1969). Ph.D. thesis, University of Melbourne, Melbourne, Australia.
Parish, C. R. (1971a). *J. Exptl. Med.* 134:1.
Parish, C. R. (1971b). *J. Exptl. Med.* 134:21.
Parish, C. R. (1971c). *Proc. N.Y. Acad. Sci.* 181:108.
Parish, C. R., and Ada, G. L. (1969a). *Biochem. J.* 113:489.
Parish, C. R., and Ada, G. L. (1969b). *Immunology* 17:153.
Parish, C. R., and Marchalonis, J. J. (1970). *Anal. Biochem.* 34:436.
Parish, C. R., Wistar, R., Jr., and Ada, G. L. (1969). *Biochem. J.* 113:501.
Pearson, M. N., and Raffel, S. (1971). *J. Exptl. Med.* 133:494.
Pinnas, J. L., and Fitch, F. W. (1966). *Internat. Arch. Allergy* 30:217.
Pye, J. (1968). *Proc. Aust. Biochem. Soc.* 1:83.
Shellam, G. R. (1969a). *Immunology* 16:45.
Shellam, G. R. (1969b). *Immunology* 17:267.
Shellam, G. R., and Nossal, G. J. V. (1968). *Immunology* 14:273.
Shortman, K., Diener, E., Russell, P., and Armstrong, W. D. (1970). *J. Exptl. Med.* 131:461.
Steward, J. P. (1969). *Fed. Proc.* 28:980.

The Transfer of Immunity with Macrophage RNA

Joel W. Goodman

Department of Microbiology
University of California, San Francisco
San Francisco, California

I. INTRODUCTION

The origin of the currently accepted doctrine that interplay between more than one type of cell is involved in the induction of the immune response can be traced to descriptions a decade ago of the transfer of immunity with RNA-rich extracts from cells exposed for short periods of time to antigen (Fishman, 1961; Fishman and Adler, 1963). This ancestry is peculiarly ironic in that the interpretation of those early experiments as a cooperative effect is more suspect now than then, albeit a considerable number of investigators have since taken up the gauntlet.

The donors of the RNA were peritoneal exudate (PE) cells, which are normally composed of 70-90% macrophages, the remainder being a mixture of small lymphocytes and polymorphonuclear leukocytes, with traces of other cell types. On the basis of these relative proportions, the observed effect was attributed to the macrophage. It is clear from a number of prior and subsequent observations that the macrophage is indeed an important, if not essential, accessory cell for the induction of the immune response to at least some antigens. Theories of macrophage function range from the nonspecific concentration and presentation of antigen to the transfer of informational RNA. Upon close examination, however, there is little, if any, persuasive evidence implicating the macrophage as a determinant of the specificity of immune responses. On the other hand, a substantial, compelling, and continually growing body of evidence portrays the humoral antibody response as a sequela of the interaction of antigen and two functionally distinct, antigen-specific lymphoid cell lines, one of which is responsible for cellular immune phenomena, whereas the other is the secretor of antibody (Talmage *et al.,* 1970).

93

I attempt this review on the premise that this may be an appropriate time to pause and take stock of the position of the macrophage, unencumbered by the sundry permutations in experimental design which have been thrust into the campaign. For that reason, I propose to concentrate on the few clearly established principles which have emerged, on the most significant questions remaining unanswered, and on whatever plausible approaches to their solution come to mind.

II. THE HEIGHTENED IMMUNOGENICITY OF ANTIGENS ASSOCIATED WITH MACROPHAGES

A. Immune Responses of Fractionated Cell Populations

Although it had been known since the time of Metchnikoff that macrophages ingest particulate (and, as it developed later, soluble) material, including immunogenic substances, it was not until mixed populations of lymphoid cells were partially resolved into their component elements that some insight into the cellular requirements for immune induction could be established with confidence. Cell populations from mice have been fractionated on the basis of selective adherence of macrophages to glass or plastic surfaces (Mosier, 1967; Mosier and Coppleson, 1968), density (Haskill *et al.,* 1970; Shortman *et al.,* 1970; Raidt *et al.,* 1968), and size (Shortman *et al.,* 1970). The separations in all cases may be considered incomplete, since it is impossible to document the virtual absence of one type of cell from a large population of another, and it is best to assume that these procedures yield enriched rather than pure populations of macrophages and lymphocytes.

Nevertheless, when the fractions were assayed for their capacity to make primary responses *in vitro* to antigens, which in most cases have been heterologous erythrocytes, certain relationships clearly emerged. Neither lymphocyte-rich nor macrophage-rich populations were capable by themselves of responding to red blood cells. On the other hand, primed lymphocytes mounted a secondary response in the absence of macrophages (Pierce, 1969). Again, it must be emphasized that this may be a quantitative rather than absolute difference, in view of the probability that the cells were imperfectly resolved.

The requirement for macrophages apparently depends to some degree on the physical nature of the antigen, since they were not needed for a primary response to polymerized bacterial flagellin (Shortman *et al.,* 1970). Although the reasons for this difference are uncertain, one obvious distinction between erythrocytes and flagellin is their dimensions, implying that degradation of particulate antigens may be an essential antecedent to immunity. However, as we shall see shortly, even soluble antigens seem to have enhanced immunogenicity when associated with macrophages. We can, therefore, safely conclude that the macrophage is at least an amplifier of the primary immune response.

B. Retention and Localization of Antigen by Macrophages

What the macrophage does with the antigen it consumes has been the subject of a number of investigations. As far as is known, about 90% of soluble antigens taken up is rapidly catabolized within lysosomes and eliminated (Unanue and Askonas, 1968; Unanue, 1969; Kölsch and Mitchison, 1968). It is extremely doubtful that this fraction plays any part in immune induction other than the possible circumvention of tolerance. The remaining 10% is retained for long periods either on or within the cell. Unanue and Cerottini (1970) have convincingly shown that at least a substantial part of it remains at the cell surface and that it is this surface-bound antigen which accounts for the heightened immunogenicity of macrophage-associated antigen (Gallily and Feldman, 1967; Mitchison, 1969; Unanue and Askonas, 1968; Unanue, 1969). If macrophages which had taken up keyhole limpet hemocyanin (KLH) were treated with anti-KLH antiserum or with trypsin, their immunogenicity in transfer experiments was largely eradicated (Unanue and Cerottini, 1970). These agents presumably act at the surface of the cell. Additionally, some of the antigen could be removed from cells by treatment with a chelating agent. Most of the recovered material was of large molecular size, some of it comparable to associated KLH.

Macrophages exposed to irradiation before antigen have an impaired immunizing activity, whereas irradiation following antigen has no effect (Mitchison, 1969). This may be correlated with the observation that while irradiated macrophages take up antigen with the facility of normal cells, they do not retain it as efficiently (Kölsch and Mitchison, 1968).

These findings suggest that the macrophage may enhance immunity simply by an efficient presentation of essentially unaltered antigen to appropriate lymphocytes. There are several satisfying facets to this interpretation. The evidence that specificity of immune responses resides in the lymphocyte is compelling, if not overwhelming (see, for example, Gowans and McGregor, 1965), and it appears indisputable that antigens are often recognized in essentially native form by cells which determine antibody specificity (Goodman, 1969; Sela, 1969).

It is unlikely, however, that the macrophage serves as a vehicle for antigen in the simplest sense (Mitchison, 1969). For one thing, viable macrophages are required for transfer of the primary response. For another, immunogenic activity is much greater in syngeneic than in allogeneic recipients. Add to this that animals recovering from paralysis may respond to macrophage-associated antigen under circumstances where they are unresponsive to free antigen, regardless of dose, and it seems inescapable that the presentation of antigen by these cells is in some way an active process.

III. RNA–ANTIGEN COMPLEXES

In contrast to a mechanism calling for the retention and presentation of unaltered antigen, the macrophage has been hypothesized to process or modify antigen in such a way as to render it more immunogenic, subsequently transferring the modified form to antibody-synthesizing cells. The evidence for this rests on a number of observations modeled upon the initial experiments involving transfer of immunity with RNA-rich extracts of peritoneal cells (Fishman, 1961; Fishman and Adler, 1963). While originally interpreted as possibly signifying the transfer of information, antigen was later demonstrated in such preparations (Friedman *et al.,* 1965; Askonas and Rhodes, 1965), obscuring the role played by RNA. The concept of a "superantigen" was advanced in which antigens become remarkably more potent when associated with RNA. A number of critical questions demanded address, however, before the phenomenon could be credited with physiological relevance. After all, linking antigens to almost anything will enhance immunogenicity, and it is likely, as we shall see shortly, that RNA regardless of source may possess an unusually potent adjuvant-like effect.

The questions of greatest moment can be formulated as follows:

1.   Is the formation of RNA-antigen complexes unique to macrophages?

2.   Does complex formation require newly synthesized or antigen-specific RNAs?

3.   Is complex formation an enzyme-dependent process?

4.   What is the relationship between molecular structure and immunopotency of the antigen and its capacity to form complexes with RNA?

5.   What is the chemical nature of the RNA–antigen bond?

6.   Do RNA-antigen complexes form within viable cells or are they a laboratory artifact?

A. Assessment of the Requirement for Macrophages

Let us first consider the question of the requirement for macrophages. Gottlieb (1968, 1969) has reported that the minute quantity of antigen associated with macrophage RNA was always concentrated in a fraction which comprised about 5% of the RNA and banded in cesium sulfate gradients at an absolutely invariant density of 1.588, whether or not cells were incubated with one of several different protein antigens. Disturbingly, the density of the major band, consisting of pure RNA, varied in different experiments between 1.620 and 1.676. It was contended that the minor band ribonucleoprotein, which disappeared after treatment with proteolytic enzymes, was unique to macrophages. This contention, however, was based on studies using analytical ultracentrifugation, which is considerably less sensitive than equilibrium density gradient centrifugation.

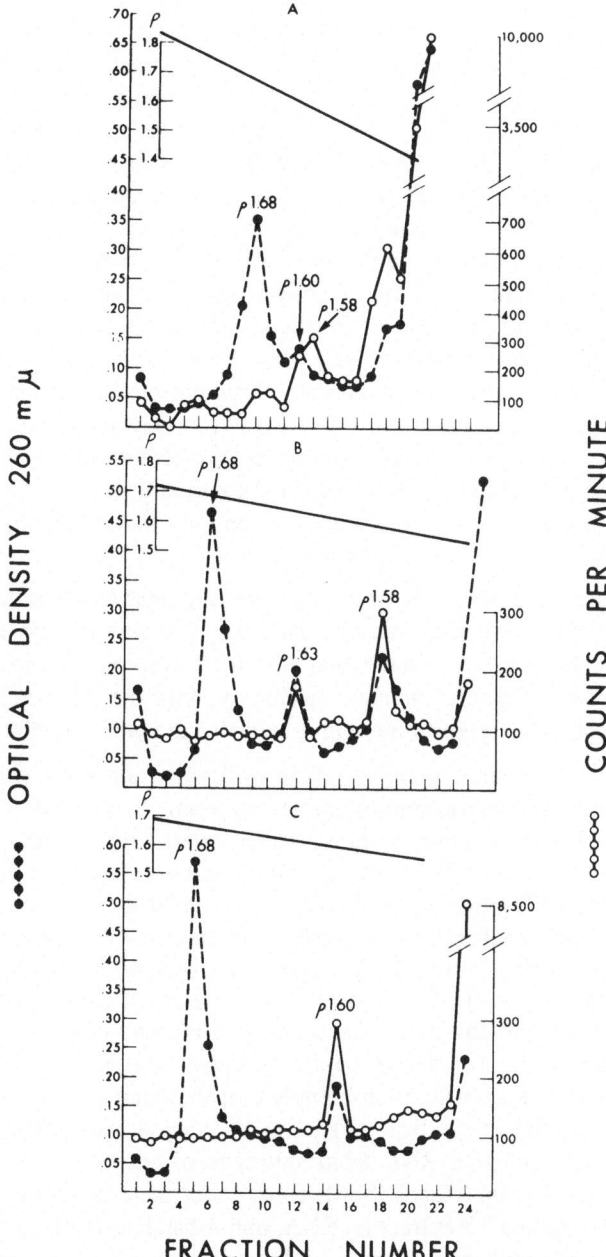

Figure 1. Equilibrium sedimentation in cesium sulfate of RNA extracted by the cold phenol method from homogenates of (A) PE cells, (B) HeLa cells, (C) *E. coli* cells incubated with TGAL-[125]I. (From Roelants *et al.*, 1971.)

Characterization of the ribonucleoprotein formed when peritoneal cells were exposed to a large synthetic copolymer of amino acids revealed that it had a molecular weight of about 12,000 and an RNA/protein ration of 3:1. Thus, the antigen component, assuming a single fragment, had a maximum molecular weight of about 3000. Other investigators have reported even smaller complexes of RNA and antigen, with total molecular weights of 1000-2000, extracted from liver, serum, or urine of immunized animals (Garvey *et al.*, 1970). These complexes were very heterogeneous and have not yet been well characterized. The role of such small fragments of antigen in the induction of the immune response is incompatible with several of the properties of antibody molecules. As mentioned earlier, many antibodies have been shown to possess specificity which is dependent on the tertiary or quaternary structure of the antigen (Goodman, 1969; Sela, 1969). In addition, the net charge of antibody molecules is affected by the net charge of the entire antigen molecule, regardless of the charge carried by the antigenic determinant (Sela, 1969). These features force the conclusion that antigens are frequently recognized in essentially native form by cells which determine antibody specificity and structure.

Of course, it is true that antibodies are also made to interior or buried determinants, indicating that antigens partially degraded *in vivo* may still be immunogenic. It might be argued that different antigens are processed along different pathways by the immune machinery, but the adoption of so convoluted a hypothesis requires more telling evidence to lend it credibility.

On the other hand, there is telling evidence against the existence of a unique macrophage ribonucleoprotein. It has been demonstrated that RNA-antigen minor bands in cesium sulfate gradients were formed with equal facility whether *Escherichia coli*, HeLa, or peritoneal cells furnished the RNA (Fig. 1) (Roelants *et al.*, 1971), thereby conflicting with conclusions reached on the basis of analytical ultracentrifugation. Moreover, the minor bands appeared at variable densities ranging from 1.56 to 1.63, whereas the bulk of the RNA banded at a reproducible density of 1.68. This would be expected if the major band represented pure RNA and the minor bands were composed of RNA associated to varying degrees with proteins or peptides. Extraction of the RNA with hot rather than cold phenol resulted in complete dissociation of the complexes and disappearance of the minor bands of lower buoyant density. The difficulty of extracting uncontaminated RNA from any type of cell by milder cold phenol treatment is well known (Mach, 1968). Thus, under the conditions used by both groups of investigators for extracting RNA, minor bands containing antigen were not unique to macrophages. There is no basis for assuming that the complexes formed with peritoneal cells were qualitatively different from those formed with other cell types.

Furthermore, while RNA had a pronounced enhancing effect in priming

Table I. Priming Activity in CSW and C3H Mice of RNA
in Complex with or Mixed with TGAL[a]

Priming injection	Mouse strain	Percent TGAL-I[131] bound by serum[b]				
		0-10	11-25	26-50	51-75	76-100
PE cell minor band RNA	CSW	14				
	C3H	6				
E. coli minor band RNA	CSW	9	1			
	C3H	6				
PE cell major band RNA (3 µg) + TGAL (1 µg)	CSW	6	1	3		
	C3H	6				
PE cell major band RNA (3 µg)	CSW	10	1			
	C3H	4				
E. coli major band RNA (3 µg) + TGAL (1 µg)	CSW	2	4	3	6	3
	C3H	6				
E. coli major band RNA (3 µg) + TGAL (0.1 µg)	CSW	6				
	C3H	6				
E. coli major band RNA (3 µg)	CSW	10	1			
	C3H	6				

[a] All priming injections were administered in saline. The normal priming injection for CSW mice consists of 10 µg of TGAL in complete Freund's adjuvant; the same quantity of polypeptide in saline does not prime. (From Roelants et. al., 1971.)

[b] Twenty-five microliters of 1:10 dilution of antiserum mixed with 0.0025 µg of TGAL.

mice for immune responses to a synthetic branched polypeptide, E. coli RNA was at least as effective as RNA from peritoneal cells (Table I) (Roelants et al., 1971). Other reports strengthen the proposition that RNA from diverse sources may exert a conspicuous adjuvant-like activity, the mechanism of which is unclear. A synthetic doubly helical RNA, polyadenylic-polyuridylic acid, transformed a strain of mice which responded poorly to another synthetic polypeptide into one which responded well (Mozes et al., 1971). The probability of a resemblance between this synthetic polynucleotide, E. coli RNA, and macrophage RNA in the sense of antigen recognition would appear to be vanishingly remote. Again, it would have to be maintained that different mechanisms were involved, and the credibility of that position is contingent on experimental evidence.

B. Assessment of the Existence of Antigen-Specific RNAs

The peritoneal cell RNA involved in complex formation did not appear to be antigen specific, based on a number of observations. RNA could not be saturated by increasing the quantity of a particular antigen within reasonable limits, and newly synthesized RNA was not required for the formation of complexes (Roelants and Goodman, 1969). These findings are consistent with DNA-RNA hybridization experiments which showed that RNA from macrophages which had been incubated with an antigen was indistinguishable from the RNA of normal macrophages (Gottlieb *et al.,* 1967; Raska and Cohen, 1968), attesting to the absence of unique or antigen-specific RNAs in this phenomenon.

C. Assessment of the Requirement for an Enzyme

The kinetics of association excluded the involvement of an enzyme (Roelants and Goodman, 1969) and, together with the dissociation of complexes by hot phenol (Roelants *et al.,* 1971), rendered unlikely covalent binding between RNA and antigen. Varying the times of incubation and the temperature, heating PE cell homogenates, treating them with pronase or dithiothreitol prior to incubation with antigen, or incubating the components in the presence of dispersing agents all exerted no apparent influence on the formation of complexes, weakening prospects that their assembly involved an active biological process.

D. Relationship Between the Capacity to Complex with RNA and Immunogenicity

Another parameter of immunological significance is the relationship between the immunogenicity, or immunopotency, of a molecule and its capacity to form complexes with RNA. In an extensive exploration of this relationship, no correlation between these properties was found (Roelants and Goodman, 1969). A γ-D-glutamyl polypeptide, whether precipitated with alum or complexed with methylated albumin, behaved indistinguishably with respect to capacity to associate with RNA. In the former form, the polypeptide is at best a very weak immunogen in rabbits, if not totally nonimmunogenic, whereas the latter form regularly elicits antipolypeptide antibody. Other examples substantiate this point. Although pure polysacharides are nonimmunogenic in rabbits, dextra-^{14}COOH gave appreciable binding with RNA while dextran-^{3}H did not. Cells from a strain of mice which responds poorly to a synthetic polypeptide yielded the same degree of association with RNA as cells from a strain which responds well. Free glutamic acid associated with RNA as efficiently as human γG immunoglobulin, while steroids were completely negative in association assays.

Egan *et al.* (1971) found that positively charged synthetic polypeptides associated with RNA from peritoneal exudate cells to a greater degree than negatively charged polymers. Neutral, but not totally uncharged, polymers gave negligible or undetectable levels of association. The polymers became associated with low molecular weight RNA, and there was no apparent difference in either the quantitative or qualitative features of the complexes whether lymph node, spleen, bone marrow, or testes cells were used. Complex formation between RNA and basic molecules is readily explicable on the basis of electrostatic interaction. Indeed, a method used to obtain antibodies against nucleic acids involved the use of nucleic acid-methylated albumin complexes as immunizing agents (Plescia *et al.*, 1964).

In another investigation, an immunogenic random copolymer of L-amino acids associated with RNA more efficiently than its weakly or nonimmunogenic D-amino acid counterpart, although an excess of the D-polymer inhibited the association of the L-compound (Gottlieb, 1971). It would have been of interest to know if the D-polymer could in fact inhibit the immune response to the L-polypeptide, but this was not reported. Thus the question of whether the differences in behavior of the two polypeptides toward RNA were due to physical properties unrelated to the immune response remained unresolved.

On balance, it can only be concluded that molecular size and structural complexity, which play such vital roles as determinants of immunogenicity, are not factors in binding with RNA. An analysis of the properties of molecuies which bind to RNA and those which do not revealed that the common denominator shared by all the binders and absent from all the nonbinders was the possession of charged groups (Roelants and Goodman, 1969). The inability of uncharged molecules to complex with RNA, within limits of detection, vitiates any contention that binding might be a necessary but insufficient condition for inducing an immune response, since uncharged molecules can be immunogenic (Goodman, 1969; Sela, 1969).

E. The Nature of the RNA-Antigen Chemical Bond

What remains, then, is the rather curious situation in which RNA, a very acidic molecule, can be noncovalently joined to other molecules which carry acidic functional groups. In place of an anticipated electrostatic repulsion, we find instead the formation of surprisingly stable complexes (Roelants and Goodman, 1968). Another intriguing piece of the puzzle is that complex formation does not occur between purified RNA and antigen, but takes place only with intact cells or crude cell homogenates (Roelants and Goodman, 1968), indicating a need for some vital cellular component.

Certain recently disclosed properties of tRNA provided important clues to the solution of the mystery, particularly since association with antigen appeared

Table II. Formation of Complexes Between Poly-γ-D-Glutamic Acid-^{3}H and Purified PE Cell RNA at Different Mg^{2+} Concentrations[a]

(Mg^{2+})[b] (M)	Radioactivity in complex[c]	
	Total (c.p.m.)	Percent
0.02	16,010	15
0.004	23,970	22
0.002	22,056	21
0.0002	22,814	21

[a] From Roelants and Goodman (1969).

[b] As $MgCl_2$, in 0.01 M tris-HCl (pH 7.4), 0.001 M EDTA, and 0.01% sodium dodecyl sulfate.

[c] Corrected for controls (antigen in the complex in the absence of Mg^{2+} < 1%).

to involve 4S RNA (Roelants and Goodman, 1968). tRNA exists in essentially two forms in the cytoplasm. A more open or expanded form, considered a precursor, can be transformed to the mature, coiled form by Mg^{2+} ions, which chelate phosphate residues in the RNA molecule (Miller and Steiner, 1966; Cramer *et al.,* 1968); the coiled form is extremely stable (Takanami *et al.,* 1961). High concentrations of Mg^{2+} ions can aggregate tRNA, presumably by the formation of intermolecular chelates (Miller and Steiner, 1966). These findings, coupled with the characteristics of RNA-antigen complex formation, suggested that chelation of metal ions by anionic groups could account for the formation of complexes, the vital cellular element being nothing more than divalent cations. This hypothesis was borne out by implementing the complexing of pure RNA and antigen with magnesium salts (Table II) and by dissociating complexes formed *in vivo* by exhaustive dialysis against buffers containing a chelating agent (Table III) (Roelants and Goodman, 1969). Dissociation occured more readily when dialysis was carried out near the melting temperature of tRNA than at lower temperatures. The effectiveness of a magnesium concentration as low as 2 x 10^{-4} M and the need for high concentrations of EDTA to abolish the effect suggest that few bonds are needed to form a complex. The remarkable stability of the complexes suggests that they may be stabilized by a conformational change analogous to the maturation of tRNA.

F. Are RNA-Antigen Complexes Laboratory Artifacts?

Thus, the formation of RNA-antigen complexes lacks specificity at the levels of the antigen, the RNA, the cell, and the mechanism of complex

Table III. Dissociation of a Poly-γ-D-Glutamic Acid-^{3}H-RNA Complex
Obtained *in Vivo*[a]

Dialysis	Buffer[b]	Temperature	Polypeptide/RNA ratio (% of starting complex)
−	1	20	100
+[c]	1	20	69
−	1	80	100
+	1	80	31
−	2	20	95
+	2	80	100

[a] From Roelants and Goodman (1969).

[b] 1: 0.02 M tris-HCl (pH 7.4), 0.01% sodium dodecyl sulfate, and 0.028 M EDTA; 2:1 without EDTA.

[c] The complexes were suspended in a 1 ml volume and dialyzed for 3 days against 250 ml of buffer, changed twice daily.

formation. Complexes can be formed by simply mixing RNA, any molecule which carries negatively charged groups, and divalent cations. All these ingredients are present within cells and are readily brought into contact by disruption of cells prior to extraction of RNA. Extracellular "antigens" are taken up by phagocytic cells and consequently are accessible for association with RNA. Complexes are formed with other kinds of cells when the cells are disrupted in the presence of "antigen." It should be stressed that there is still no convincing evidence that complexes form naturally within viable cells. On the basis of what evidence is available, they could well be due entirely to laboratory artifact produced by scrambling the cell's exquisite organization.

While macrophages and lymphocytes have been photographed in intimate contact, in some cases with cytoplasmic bridges between them (Schoenberg *et al.*, 1964), the reality of such features, as well as whether anything is being passed between the cells, remains conjectural.

IV. THE TRANSFER OF CELLULAR IMMUNITY WITH RNA

The transfer of cellular immunity to naive lymphocytes by RNA extracted from lymph nodes of immune animals has been allegedly achieved using several different experimental models. These include the rejection of skin and tumor allografts (Mannick and Egdahl, 1962; Alexander *et al.*, 1967); the release of migration inhibitory factor (MIF) (Thor and Dray, 1968a,b; Jureziz *et al.*, 1968), a substance released by sensitive lymphocytes in the presence of specific antigens which inhibits the movement of macrophages; the capacity to give a delayed cutaneous reaction to antigen *in vivo* (Jureziz *et al.,* 1970); and the ability of lymphoid cells to synthesize DNA in response to antigen (Thor and Schlossman, 1969). Since lymph nodes rather than peritoneal exudates served as the source of RNA in most of these studies, lymphocytes rather than macrophages constituted the predominant cell type.

In essence, the protocols have been similar to those used to transfer humoral immunity, and the same qualifications apply to interpretation of the results. A difference which may prove significant is that RNA was extracted from cells derived from immune animals rather than from cells which had been directly exposed to antigen *in vitro* immediately before preparation of the RNA. This RNA was presented to lymph node cells from normal donors, which in turn were either confronted directly with antigen or transferred to normal recipients, which were then challenged with antigen.

It might be expected that residual antigen in lymph nodes from immune animals would be substantially less than in freshly exposed peritoneal cells, but the presence of sufficient antigen to exert a "superantigen" effect has not been excluded. Control experiments showed that neither antigen alone nor RNA from nonimmune animals was able to convert lymphocytes into sensitized cells. Some of the active extracts were analyzed and contained no detectable DNA but had small amounts of protein. They were inactivated by ribonuclease but not by deoxyribonuclease or trypsin. These observations indicate that RNA was essential for the effect, but resistance to trypsin is not an adequate criterion for the exclusion of antigen. For one thing, many proteins are not degraded by trypsin, and, for another, the RNA may have protected antigen from attack by the enzyme. Little is known as yet about the mechanism involved in the induction of cellular immunity by RNA, but at this point there seems little reason to classify it differently from the induction of humoral immunity. It would be useful to determine if RNA from normal animals can be converted to an active product by chelation with antigens. This could be accomplished most efficiently by mixing the two components with divalent cations at or near the melting temperature of RNA.

It is worth noting that the accumulated reports of this phenomenon derive from relatively few laboratories. Investigators are reluctant to publish negative findings since many factors can be responsible for lack of success. It would be surprising, however, if other investigators have not attempted similar experiments. One negative report has been published recently in which RNA from the lymph nodes of strain 2 guinea pigs, which make an immune response to poly-L-lysine (PLL) and had been immunized with DNP-PLL, was added to cells from normal nonresponder strain 13 animals. The strain 13 cells were then assayed for their ability to be stimulated to proliferate by the antigen and failed to respond (Bluestein *et al.,* 1970). These results directly contrast with those of other investigators using an identical experimental model (Thor and Schlossman,

Table IV. Identification of the Allotypic Specificity of the γM Antibodies in the Plaques by Inhibition of the Appearance of the Plaques with Antiallotype Antibodies Incorporated into the Agarose[a]

Source of spleen cells (SpC)	Expt. No.	1. Genotype of the SpC used for plaque assay	2. Genotype of donor of "immune" RNA	3. No serum or antibody added	4. Normal rabbit serum (0.2 ml)	Antiallotype antisera (0.2 ml)[b]		γG immunoglobulin (100 μg[c])	
						5. Anti-b4	6. Anti-b5	7. Anti-b4	8. Anti-b5
Immunized rabbits	1	b^4b^4	None	340	325	4	315	10	319
	2	b^4b^4	None	382	365	1	358	1	351
	3	b^5b^5	None	265	250	245	2	219	8
	4	b^5b^5	None	305	292	302	3	286	9
Nonimmunized rabbits	5	b^4b^4	None	7	5	3	1	7	5
	6	b^5b^5	None	2	4	5	0	8	5
	7	b^4b^4	b^4b^4	365	339	0	318	0	323
	8	b^4b^4	b^5b^5	270	247	250	2	240	4
	9	b^5b^5	b^4b^4	352	331	0	303	1	309
	10	b^5b^5	b^5b^5	235	216	212	4	220	3
	11	b^4b^4	b^4b^4	361	331	8	315	3	324
	12[d]		b^5b^5	273	220	205	5	212	0
	13[e]	b^5b^5	b^4b^4	335	312	9	296	5	293
	14[e]		b^5b^5	251	240	233	9	236	7

[a] Data are expressed as PFC/10⁶SpC. (From Bell and Dray, 1969.)
[b] Amount of antiallotype antisera incorporated into the agarose mixture (2.2 ml total volume).
[c] Amount of γG immunoglobulin in the agarose mixture (2.2 ml total volume).
[d] Same donor of SpC as in Expt. 11.
[e] Same donor of SpC as in Expt. 13.

1969). However, the positive results have only been reported in abstract form, making it impossible to compare experimental details.

V. INFORMATIONAL RNA

To this point, the phenomena under consideration can be accounted for, at least hypothetically, by the existence of an abnormally active form of antigen. The presence of antigen has been clearly demonstrated in some of the RNA preparations and has not been rigorously excluded from any. We shall now consider an entirely different set of observations which are not explicable on the basis of a "superantigen" but require a more specific role for the RNA.

In these experiments, RNA extracted from peritoneal exudate or lymph node cells of immune rabbits homozygous for a given allotype was either mixed with normal cells from, or injected into, animals homozygous for the allelic allotype. A portion of the antibody produced by the recipients carried donor allotypic markers. Allotypic specificity was assessed either by neutralization of bacteriophage (Adler *et al.*, 1966) or by localized hemolysis in gel (Bell and Dray, 1969, 1970, 1971), depending on whether phage particles or erythrocytes served as antigen. Most or all of the IgM antibody appeared to bear donor allotypic markers, and in the most recent reports some of the IgG was also of donor type. The allotypic specificity of antibody produced by plaque-forming cells was identified either by direct precipitation within the plaque or by inhibition of plaque formation, using antisera to the b4 and b5 allotypic specificities of immunoglobulin light chains.

The quantitative data are of importance in establishing the veracity of these findings, particularly since much of the earlier work with RNA was based on rather marginal quantitative data. However, the values in Table IV leave little doubt of the reality of the effect, since the plaque count is two orders of magnitude higher in cells treated with RNA, and all of the plaque can be inhibited with antiserum specific for the allotype of the RNA donor.

The interpretation of the results is another matter and remains far from settled. As already mentioned, it is difficult to conceive of a mechanism by which antigen itself could induce the expression of foreign genetic markers; consequently, the RNA must be exerting a highly specific effect. The most direct and obvious role for RNA in the transfer of genetic information would be to serve as a messenger for the translation of immunoglobulin polypeptide chains. Such a proposal, however, is faced with a number of difficulties.

If the activity of these RNA preparations is indeed due to a messenger function, then this activity should be demonstrable in a totally cell-free system. Translation of specific messenger RNAs from eukaryotic cells has been achieved *in vitro* in a few instances, although the RNAs were extracted with sodium dodecyl sulfate rather than with the more stringent phenol treatment used for the preparations in question here. The α and β chains of rabbit hemoglobin have been synthesized in a cell-free system derived from *E. coli* (Laycock and Hunt, 1969), and mouse β chain has been fabricated in a system derived from rabbit reticulocytes (Lockard and Lingrel, 1969). Of much greater relevance here, the translation by a rabbit reticulocyte system of RNA extracted from a mouse plasma cell tumor which synthesized and secreted an immunoglobulin light chain has recently been reported (Stavnezer and Huang, 1971). The polypeptide chains synthesized *in vivo* and *in vitro* were compared by differential isotope labeling and tryptic peptide profile analysis. They matched almost perfectly. If under similar circumstances allotypic markers characteristic of the donor of the RNA could be shown to be synthesized, it would provide compelling evidence for the transfer of messenger RNA in the experimental system. In addition, actinomycin D should be used to distinguish between direct translation and a pathway which proceeds through DNA and RNA synthesis prior to translation.

Another impediment to attributing the observed results to messenger activity is the clearly established fact that the heavy and light chains of immunoglobulins produced by a particular clone of cells are almost uniquely compatible (Roholt *et al.*, 1965; Zappacosta and Nisonoff, 1968), implying that the selection of variable region genes for paired chains is coupled in some way, albeit the genes for the chains are unlinked. The RNA used to transfer immunity has invariably been taken from mixed populations of cells, so random combinations of messenger RNAs for heavy and light chains would be expected to result almost exclusively in poorly matched or inefficient antibody molecules.

It is probably pointless to attempt to assign numbers to these considerations, since many variables are involved. For example, we do not know the limiting avidity of antibody required for lytic activity and resultant plaque formation or the quantity of RNA taken up by lymphoid cells. However, it is clear that multiple copies of messenger RNAs sufficiently matched to produce a reasonably functional antibody molecule must enter each cell which forms a discernible plaque. This raises another interesting question. Since it is extremely unlikely that identical messengers would enter any particular cell in multiple proportions, it would be expected that each cell transformed by RNA might produce a heterogeneous population of antibody molecules, reflecting the heterogeneity of the absorbed RNA. This would be very different from the normal situation, in which there is good reason to believe that a given cell is producing a very homogeneous product.

Alternatively, the RNA need not directly act as messenger but could conceivably serve as a template for a RNA-dependent DNA polymerase. In this scheme, RNA would be transcribed to DNA, which could then serve to turn out multiple copies of the RNA, which in turn would be translated into protein. Thus, a single RNA molecule could give rise to many duplicates of itself. There is precedent for a mechanism of this sort. First found in cells infected with RNA tumor viruses, RNA-dependent DNA polymerase activity has recently been demonstrated in normal human lymphoid cells (Scolnick *et al.*, 1971). Thus far, RNA-DNA hybrid molecules have been the most effective templates, with DNA transcripts of the RNA strand being synthesized. Nevertheless, these findings offer an intriguing explanation for the transfer of genetic information by RNA which in several important respects is more satisfying than the hypothesis that the transferred RNA serves directly as messenger. Exploration of this mechanism is clearly warranted and should be feasible in an *in vitro* system. If immunoglobulin of donor type was synthesized in the absence of actinomycin D but not in its presence, the possibility that a RNA-dependent DNA polymerase was involved would be materially strengthened.

A third possibility to account for the experimental observations is that the structural genes for the allotypically different polypeptide chains may not be truly allelic (Bell and Dray, 1971). While the extensive body of genetic data is clearly consistent with allelism, genes for both the b4 and b5 light chains might indeed be present in animals homozygous for each type, but only one might be expressed due to genes which control the synthesis of the polypeptide chains. Thus, the allelic genes might be control rather than structural genes. If this were so, the RNA extracts might exert their effect by modifying the action of the control genes.

Whatever the mechanism of this fascinating phenomenon, caution must be exercised in extrapolating a highly artificial experimental system to the natural state of the organism. Although the observations made with RNA extracts are provocative and of interest in their own right, they in fact do not constitute proof that the induction of the immune response *au naturel* requires or involves the transfer of RNA from one cell to another. We can liken this to the synthesis of rabbit hemoglobin by an *in vitro* system derived from *E. coli*. It would be absurd to conclude on the basis of these results that *E. coli* normally manufactures hemoglobin.

VI. CONCLUSIONS

Although it is clear from a number of observations that the macrophage plays an important, and indeed perhaps essential, part in the immune response to many antigens, precisely how it plays its part is still a matter of controversy. When all the available information is carefully weighed, the most satisfying mechanism that emerges is one in which the macrophage concentrates and retains antigen on its surface, providing an effective locus for the interaction of immunocompetent cells with antigen and with each other. If macrophage RNA is a specific ingredient in the induction of immunity, it has yet to be convincingly demonstrated. Essentially all the properties of macrophage RNA which were considered significant and unique are, in fact, demonstrable with RNA from other sources. In particular, any antigen which bears anionic groups can complex with RNA regardless of source, and such complexes are more immunopotent than free antigen.

The transfer of information with RNA extracts has thus far been reported by very few investigators and sorely needs independent confirmation. For the present, it may be attributed to lymphoid cells, which comprise a minor population in peritoneal exudates and the major cell type(s) in lymphoid tissue. The phenomenon cannot readily be accounted for by a more potent form of antigen but requires a specific role for RNA. This role could conceivably take the form of messenger RNA, a template for a RNA-dependent DNA polymerase, or a modifier of a gene involved in the control of immunoglobulin synthesis. The messenger hypothesis poses the greatest difficulties, but if extracted RNA proved to be active in a completely cell-free system for protein synthesis in the presence of actinomycin D, then the evidence in its favor would be compelling.

In any event, the experimental systems employed thus far cannot establish physiological relevance for a mechanism of immune induction involving the intercellular transfer of RNA.

ACKNOWLEDGMENT

This study was supported in part by United States Public Health Service Grants AI 05664, AM 08527, and AI 00299.

REFERENCES

Adams, A., Lindahl, T., and Fresco, J. R. (1967). *Proc. Natl. Acad. Sci.* 55:1684.
Adler, F. L., Fishman, M., and Dray, S. (1966). *J. Immunol.* 97:554.
Alexander, P., DeLorme, E. J., Hamilton, L. D. G., and Hall, J. G. (1967). *Nature (Lond.)* 213:569.
Askonas, B. A., and Rhodes, J. M. (1965). *Nature (Lond.)* 205:470.
Bell, C., and Dray, S. (1969). *J. Immunol.* 103:1196.
Bell, C., and Dray, S. (1970). *J. Immunol.* 105:541.
Bell, C., and Dray, S. (1971). *Science* 171:199.
Bluestein, H. G., Green, I., and Benacerraf, B. (1970). *Proc. Soc. Exptl. Biol. Med.* 135:146.
Burdon, R. H. (1967). *J. Mol. Biol.* 30:571.
Cramer, F., Doepner, H., Haar, F. V. D., and Schlimme, E. (1968). *Proc. Natl. Acad. Sci.* 61:1384.
Egan, M. L., Smyth, R. D., and Maurer, P. H. (1971). *J. Immunol.* 107:540.
Fishman, M. (1961). *J. Exptl. Med.* 114:837.
Fishman, M., and Adler, F. L. (1963). *J. Exptl. Med.* 117:595.
Friedman, H. P., Stavitsky, A. B., and Solomon, J. M. (1965). *Science* 149:1106.
Gallily, R., and Feldman, M. (1967). *Immunology* 12:197.

Garvey, J. S., Kinderknecht, H., and Weliky, B. (1970). *Fed. Proc.* **29**:376.
Goodman, J. W. (1969). *Immunochemistry* **6**:139.
Gottlieb, A. A. (1968). In Plescia, O. J., and Braun, W. (eds.), *Nucleic Acids in Immunology,* Springer-Verlag, New York, p. 471.
Gottlieb, A. A. (1969). *Biochemistry* **8**:2111.
Gottlieb, A. A. (1971). In Beers, R. E., and Braun, W. (eds.), *Biological Effects of Polynucleotides,* Springer-Verlag, New York, p. 293.
Gottlieb, A. A., Glisin, V. R., and Doty, P. (1967). *Proc. Natl. Acad. Sci.* **57**:1949.
Gowans, J. C., and McGregor, D. D. (1965). *Progr. Allergy* **9**:1.
Haskill, J. S., Byrt, P., and Marbrook, J. (1970). *J. Exptl. Med.* **131**:57.
Jureziz, R. E., Thor, D. E., and Dray, S. (1968). *J. Immunol.* **101**:823.
Jureziz, R. E., Thor, D. E., and Dray, S. (1970). *J. Immunol.* **105**:1313.
Kölsch, E., and Mitchison, N. A. (1968). *J. Exptl. Med.* **128**:1059.
Laycock, D. G., and Hunt, J. A. (1969). *Nature (Lond.)* **222**:1118.
Lindahl, T., Adams, A., and Fresco, J. R. (1966). *Proc. Natl. Acad. Sci.* **55**:941.
Liu, W. I., and Wang, P. T. (1964). *Sci Sim.* **13**:463.
Lockard, R., and Lingrel, J. B. (1969). *Biochem. Biophys. Res. Commun.* **37**:204.
Mach, B. (1968). In Williams, C. A., and Chase, M. W. (eds.), *Methods in Immunology and Immunochemistry,* Vol. II, Academic Press, New York, p. 328.
Mannick, J. A., and Egdahl, R. N. (1962). *Ann. Surg.* **156**:356.
Miller, D. B., and Steiner, R. F. (1966). *Biochemistry* **5**:2289.
Mitchison, N. A. (1969). *Immunology* **16**:1.
Mosier, D. E. (1967). *Science* **158**:1575.
Mosier, D. E., and Coppleson, L. W. (1968). *Proc. Natl. Acad. Sci.* **61**:542.
Mozes, E., Shearer, G. M., Sela, M., and Braun, W. (1971). In Beers, R. E., and Braun, W. (eds.), *Biological Effects of Polynucleotides,* Springer-Verlag, New York, p. 197.
Pierce, C. W. (1969). *J. Exptl. Med.* **130**:345.
Plescia, O. J., Braun, W., and Palczuk, N. C. (1964). *Proc. Natl. Acad. Sci.* **52**:279.
Raidt, D. J., Mishell, R. I., and Dutton, R. W. (1968). *J. Exptl. Med.* **128**:681.
Raska, K., and Cohen, E. P. (1968). *Nature (Lond.)* **217**:5130.
Roelants, G. E., and Goodman, J. W. (1968). *Biochemistry* **7**:1432.
Roelants, G. E., and Goodman, J. W. (1969). *J. Exptl. Med.* **130**:557.
Roelants, G. E., Goodman, J. W., and McDevitt, H. O. (1970). *J. Immunol.* **106**:1222.
Roholt, O. A., Radzimski, G., and Pressman, D. (1965). *J. Exptl. Med.* **122**:785.
Sabbadini, E., and Sehon, A. H. (1967). *Internat. Arch. Allergy* **32**:55.
Schoenberg, M. D., Mumaw, W. R., Moore, R. D., and Weisberger, A. S. (1964). *Science* **143**:964.
Scolnick, E. M., Aaronson, S. A., Todaro, G. J., and Parks, W. P. (1971). *Nature (Lond.)* **229**:318.
Sela, M. (1969). *Science* **166**:1365.
Shortman, K., Diener, E., Russell, P., and Armstrong, W. D. (1970). *J. Exptl. Med.* **131**:461.
Stavnezer, J., and Huang, R. C. C. (1971). *Nature (Lond.)* **230**:172.
Takanami, M., Okamoto, T., and Watanabe, I. (1961). *J. Mol. Biol.* **3**:476.
Talmage, D. W., Radovich, J., and Hemmingsen, H. (1970). *Advan. Immunol.* **12**:271.
Thor, D. E., and Dray, S. (1968*a*). *J. Immunol.* **101**:51.
Thor, D. E., and Dray, S. (1968*b*). *J. Immunol.* **101**:469.
Thor, D. E., and Schlossman, S. (1969). *Fed. Proc.* **28**:629.
Unanue, E. R. (1969). *J. Immunol.* **102**:893.
Unanue, E. R., and Askonas, B. A. (1968). *J. Exptl. Med.* **127**:915.
Unanue, E. R., and Cerottini, J. C. (1970). *J. Exptl. Med.* **131**:711.
Zappacosta, F., and Nisonoff, A. (1968). *J. Immunol.* **100**:781

Relationship of Events at the Lymphocyte Cell Surface to Gene Expression: Approaches to the Problem

Richard A. Lerner

Department of Experimental Pathology
Scripps Clinic and Research Foundation
La Jolla, California

I. INTRODUCTION

One of the central issues in biology is to understand how events at cell surfaces alter gene expression. Some examples of this in prokaryotic and eukaryotic cells are (1) transfer of sex factor in bacteria; (2) effects of colicins in bacteria; (3) contact inhibition; (4) fertilization of the egg; (5) effects of certain hormones; and (6) induction of an immune response. Although these examples seem diverse, the questions to be approached are similar. It is essential to understand the nature of surface receptors, how the "linkage" between the surface and cellular genes occurs, and what ultimate alterations in macromolecular synthesis follow an event at the cell's surface.

Among the eukaryotic systems, the investigator has an advantage when studying the immune system because the nature of the receptor and event at the plasma membrane which lead to altered gene expression and cell proliferation are known. It now seems certain that plasma membrane-associated immunoglobulin (M-Ig) is the receptor and union of this immunoglobulin (Ig) with antigen is the event which initiates cellular proliferation and differentiation.

An important technical advance for study of the problems outlined above in the immune system was the establishment of continuously growing diploid lymphocytes. These cells can be thought of as being arrested in differentiation

*Recipient of NIH Career Development Award AI-46372.

somewhere between the "G_0" lymphocyte and the plasma cell, and both secrete Ig and insert it into their plasma membranes (see below).

We have been studying continuous cultures of diploid lymphocytes as a model system in which some of the questions outlined above can be approached. It is the purpose of this report to review some aspects of our progress. The areas with which we shall deal are (a) general properties of diploid lymphocytes in culture; (b) quantitative aspects of M-Ig, (c) control of synthesis of M-Ig; (d) molecular events during the rest to proliferation transition in lymphocytes; and (e) beginning observations on plasma membrane-associated macromolecules which may be involved in the "linkage" between the surface and gene expression. In keeping with the intent of this series of topics in immunochemistry, no exhaustive attempt to review the literature has been made. Rather, we confine ourselves largely to data obtained in our laboratory and try to place it into biological perspective with what is known about the induction of an immune response *in vivo*.

II. GENERAL PROPERTIES OF CONTINUOUSLY GROWING CULTURED HUMAN LYMPHOCYTES

Lymphocytes for continuous culture can be obtained from the blood or splenic tissue of normal or diseased patients. In some laboratories, a success rate of approximately 50% has been obtained in establishing lymphocyte clones (Choi and Bloom, 1970). The majority of the cell lines studied have generation times of approximately 24-36 hr. The WIL_2 line has been carried in our laboratory for about 1000 passages and has remained diploid (Fig. 1). These cells differ from most other human diploid cells in that they can be cultured indefinitely without loss of viability and can be conveniently grown to a concentration of 2-3 x 10^6 cells per milliliter. Whether or not the continued growth of these cell lines is related to expression of a complete or incomplete viral genome is at present unresolved. It is clear, however, that at least one diploid cell line (WIL_2) from a normal individual had no serological or morphological evidence of any viral genome even after approximately 1000 generations in tissue culture.

The importance of a relationship between the stage of differentiation of the diploid lymphocytes in continuous culture and comparable cells after induction of an immune response *in vivo* is obvious. If one considers that the spectrum of lymphocyte differentiation covers a span between the "G_0" resting peripheral lymphocyte and the plasma cell, cultured lymphocytes morphologically are closest in the "G_0" lymphocyte (Fig. 2) and have relatively few membrane-bound polyribosomes, lysosomes, or mitochondria. The "patchy" distribution of Ig on the surface of our cultured lymphocytes as studied by

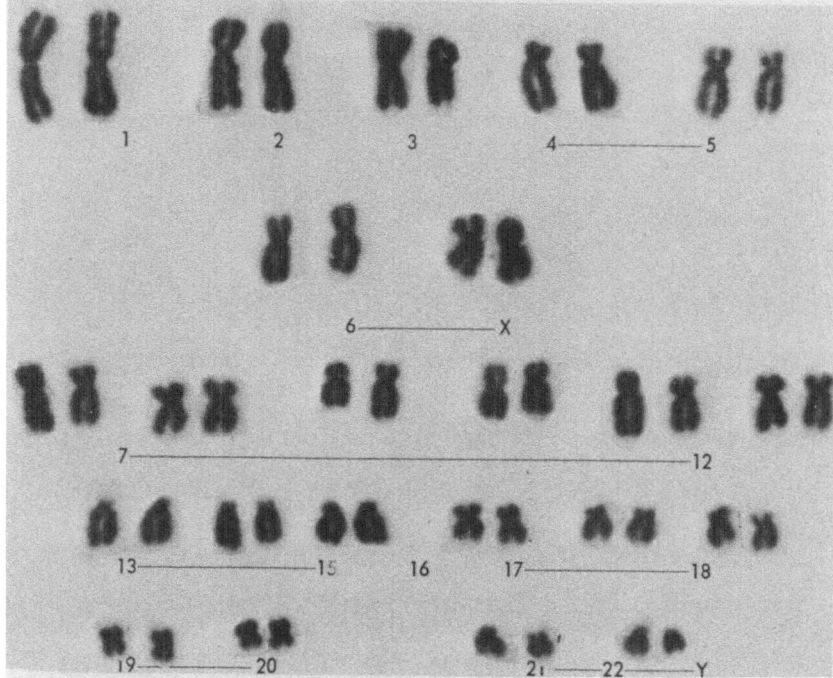

Figure 1. Karyotype of WIL$_2$ cells.

immunofluorescence was similar to that seen in human peripheral blood mono-
nuclear leukocytes stained in suspension (Dixon and McConahey, unpublished
observations). As continuous cultures capable of responding to specific antigens
become available, it may be possible to "drive" the cultured cells to differentiate, and we can then answer with certainty some of these unresolved
questions.

III. CONTROL OF SYNTHESIS OF
MEMBRANE-ASSOCIATED IMMUNOGLOBULIN

A. Quantitative Aspects

In our laboratory, a method for determination of total cellular and plasma
membrane Ig has been developed (Lerner *et al.*, 1971a). The method is based on
the Farr antigen-binding capacity test as modified for IgG fragments by Cerottini
(1968). The percent inhibition by intact cells or cytoplasmic extracts of the

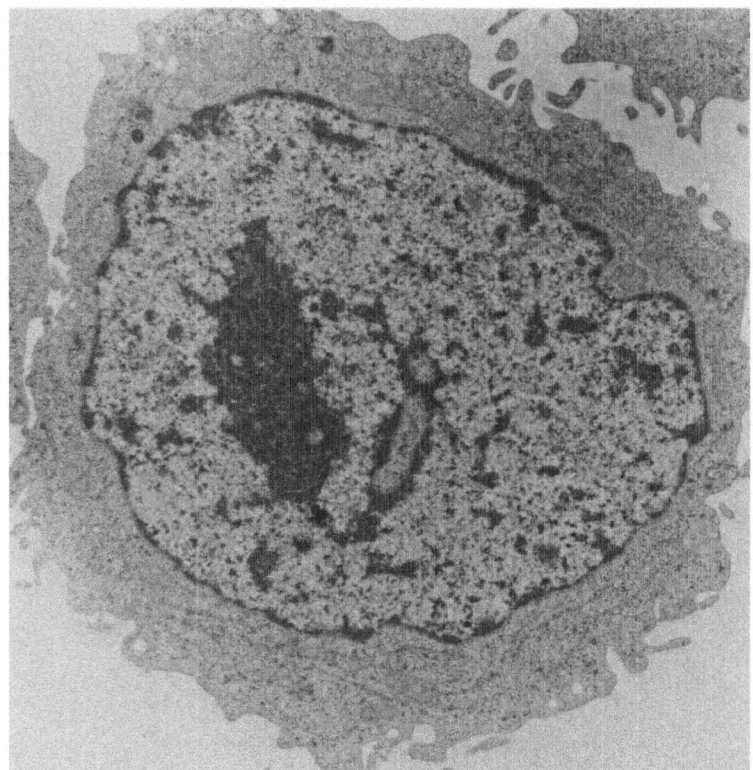

Figure 2. Electron photomicrograph of a WIL$_2$ cell. ca. × 20,000 (reduced for reproduction 50%). (Courtesy of Dr. Joseph Feldman.)

capacity of rabbit antihuman IgG antibody to combine with labeled κ and λ chains and Fc fragments is determined. Quantitation of as little as 1.0×10^{-8} g of κ and λ chains and Fc fragments is possible. Data are expressed as IgG equivalents, which relate the ability of cells and a pool of human IgG to block binding of a standard antibody to any of the trace-labeled IgM fragments studied (Lerner *et al.*, 1971*a*).

In order to quantitate the inhibition achieved by unknown materials suspected of containing IgG, it is necessary to determine the inhibition achieved by inhibitors (e.g., whole IgG molecules or fragments of IgG). We use whole IgG molecules from normal serum pools as standard inhibitors due to their relatively constant antigenicity and express the inhibition achieved by unknowns as equivalent to a certain amount of a standard IgG preparation. In this procedure, different pools of anti-IgG may give somewhat different control inhibition curves, but since the assay is based on comparative degrees of inhibition of known and unknown inhibitors, identical results are achieved with a variety of anti-IgGs as long as a single standard inhibitor is used throughout. Since different IgG blocking preparations, even those made from serum pools, will vary antigenically, their antigenic similarity to the labeled IgG fragment used will also vary. This variation is even greater when the blocking IgGs are derived from single myeloma sera. Therefore, the amount of IgG needed to achieve 50% inhibition of any reaction will depend upon the particular blocking preparation. Thus, identical data cannot necessarily be generated using different standard inhibitors.

In Table I, the membrane-associated and total IgG equivalents for a number of uncloned cells and normal human peripheral lymphocytes are listed. There was considerable variation in the nature and amount of κ and λ chains and Fc fragments among different cell lines. In one (RAJi), no M-Ig was detected. In two cell lines (1M-10 and 4265), there were excess Fc fragments in the entire cell but not excess on the surfaces of the cells. If only intact IgG molecules were bound to the surface membrane and all determinants of an IgG molecule on the cell membrane were equally available to the antisera, then the ratio of IgG equivalents determined with light chains and Fc fragments should be 1. The fact that there was always an excess of κ chains on the surface suggests that either the κ portions of intact IgG molecules were preferentially exposed or that an excess of free light chains could be held on the plasma membrane independent of the intact IgG molecule. Since Fc-μ fragments were not measured in the present study, it remains possible that some membrane-associated light chains were associated with IgM molecules and, therefore, were not free. Also, we cannot exclude the possibility that a portion of the Fc fragment was "buried" in the membrane and thus not available to antisera. In this regard, it is of interest that in two cell lines (1M-10 and 4265) having an excess of Fc fragments in the cells, an excess of κ chains was still associated with the plasma membrane. The ratio of intracellular to membrane-associated light chains varied between 15:1 and 30:1.

One difficulty with all previous attempts to measure M-Ig was that absorption of Ig from serum onto plasma membranes could not be ruled out. In our laboratory, studies were done to eliminate the possibility that M-Ig simply represented secreted Ig molecules readsorbed to the plasma membrane. WIL$_2$ and 8866 cells, which have respective phenotypes for M-Ig of κ+Fc- and κ+Fc+, were grown for 2 days, harvested, and resuspended for 24 hr in medium harvested from the other cell line (e.g., WIL$_2$ cells in 8866 medium). The phenotype of M-Ig remained the same for each cell line; thus adsorption of secreted molecules onto plasma membranes seemed unlikely (Lerner et al., 1971a).

In order to determine the degree of variability of Ig synthesis among the cultured cells, clones were obtained and studied. Twenty-one clones of WIL$_2$ and 8866 were studied, and none were found which did not synthesize M-Ig.

Table I. Equivalent Grams × 10^{-9} IgG per 10^6 Cells

	Membrane				$Total^a$				Total/membrane		
	κ chain	λ chain	Fc	Total chain/Fc	κ chain	λ chain	Fc	Light chain/Fc	κ chain	λ chain	Fc
WIL$_2$	5.8	Neg	Neg	–	186	Neg	Neg	–	32	–	–
8866	33.8	Neg	12.9	2.6	1504	Neg	388	3.9	44	–	30
1M1	31.1	Neg	Neg	–	963	Neg	Neg	–	31	–	–
1M-10	1.4	Neg	0.9	1.5	32	Neg	78	0.4	23	–	84
RAJi	Neg	Neg	Neg	Neg	Neg	Neg	Neg	–	–	–	–
4265	5.1	Neg	2.3	2.2	77	Neg	96	0.8	15	–	42
SS-5	Not done	Not done	7.9	–	Neg	1010	174	5.8	–	–	22
1788	Neg	76	Neg	–	Neg	1291	Neg	–	–	17	–

a Total = membrane + cytoplasmic immunoglobulin.

Table II. Equivalent Grams x 10^{-9} IgG per 10^6 Cloned Cells

Clone		Membrane			Total			Total membrane	
		κ chain	Fc	κ/Fc	κ chain	Fc	κ/Fc	κ chain	Fc
WIL$_2$	2A	4.6	neg	–	88	neg	–	19	–
	3A	5.7	neg	–	121	neg	–	21	–
	4A	4.9	neg	–	108	neg	–	22	–
	5A	5.8	neg	–	76	neg	–	13	–
	6A	4.1	neg	–	59	neg	–	14	–
	7A	6.8	neg	–	151	neg	–	22	–
8866	1A	68	24	2.4	1571	1083	1.5	23	46
	2A	69	27	2.9	2289	1039	2.2	33	39
	3A	38	10	3.6	1341	223	6.0	35	21
	4A	28	11	2.6	457	161	2.8	17	15
	5A	71	25	2.8	1852	503	3.7	26	20
	6A	53	22	2.4	2253	438	5.1	43	20
	7A	27	11	2.5	581	241	2.4	21	23
	8A	50	34	1.5	1239	717	1.7	25	21
	9A	61	33	1.8	902	368	2.5	15	11
	10A	56	14	4.0	559	259	2.2	10	18
	11A	68	15	4.6	963	566	1.7	14	39
	12A	55	15	3.6	439	307	1.4	8	20
	13A	52	20	2.6	644	302	2.1	12	15
	14A	90	39	2.3	1489	730	2.0	17	19
	15A	51	24	2.1	1776	1015	1.7	35	42

This strongly suggests that every cell in these two continuously growing populations was capable of synthesizing M-Ig and supports the contention that those cells found negative by fluorescent microscopy were simply in a phase of the cell cycle at which Ig was not synthesized (Buell and Fahey, 1969; Lerner and Hodge, 1971). There was considerable variation among different clones in the absolute amounts of total Ig and M-Ig (Table II).

Differences were also noted in the ratios of κ to Fc and total cytoplasm to membrane Ig fragments. The fact that ratios of membrane-associated κ chains to Fc fragments in 8866 clones were not always the same as total cytoplasm ratios suggests that IgG molecules associated with the plasma membrane in a specific way and were not simply a random percent of total cellular IgG. These differences were not owing to errors in the method since, when individual clones were restudied, the amount of Ig fragments varied by only $\pm 10\%$. In the 8866 clones where both κ and Fc fragments were present, there was always an excess of κ chain exposed on the membrane and in the cell.

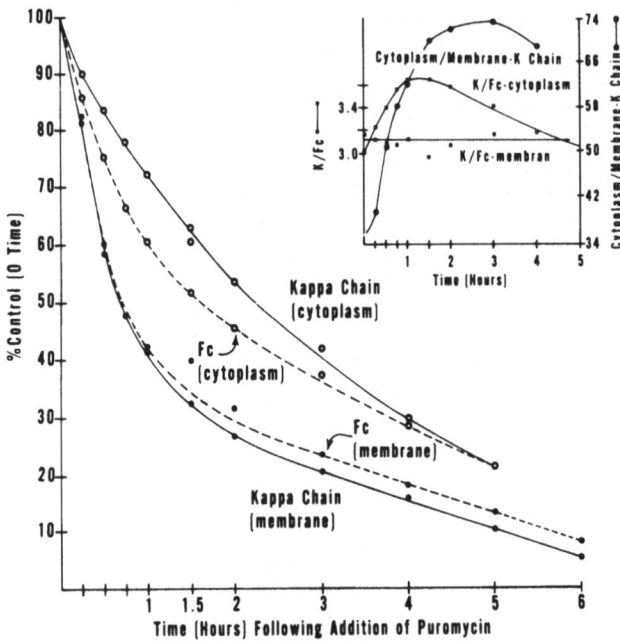

Figure 3. The amount of M-Ig and cytoplasmic Ig was determined for a culture of logarithmically growing cells (0 time), and then puromycin was added to a final concentration of 50 μg/ml. At intervals after addition of puromycin, M-Ig and cytoplasmic Ig were determined. o———o κ-chain cytoplasm; o-------o Fc-γ fragment cytoplasm; •———• κ-chain membrane; •-------• Fc-γ fragment membrane. Insert: o———o cytoplasmic/membrane κ-chain ratio; • κ-chain/Fc-γ fragment ratio in cytoplasm; ▪ κ-chain/Fc-γ-chain membrane.

The amount of IgG found on these lymphocyte surfaces may be compared to the amounts of specific antigen receptors in other systems. A clone of WIL_2 cells containing 4.6×10^{-9} g IgG equivalents on the surfaces of 10^6 cells would have 1.8×10^4 IgG molecules per cell surface. This calculation agrees with those of Byrt and Ada (1969), who determined that a single reactive lymphocyte could bind 1.7×10^4 molecules of flagellar antigen.

B. Half-Disappearance of M-Ig in Logarithmically Growing Cells Treated with Inhibitors of Protein Synthesis

To determine the half-disappearance time of M-Ig, cells were treated with puromycin to inhibit protein synthesis, and at intervals the amount of κ chain and Fc fragment present on the membrane and in the cytoplasm was determined (Fig. 3) (Lerner *et al.*, 1972).

The half-disappearance time for membrane-bound κ chains and Fc fragments was approximately 45 min, and the rates of disappearance for both Ig fragments were approximately equal (Fig. 3). The equal rates of disappearance were reflected in the constant κ/Fc ratio on the membrane throughout the study (insert, Fig. 3). The half-disappearance times for cytoplasmic κ chains and Fc fragments were approximately 2 and 1.5 hr, respectively. The faster rate of disappearance for M-Ig than for cytoplasmic Ig was emphasized by the approximate two-fold increase in the cytoplasm/membrane κ-chain ratio in the first hour after puromycin treatment. In contrast to M-Ig κ and Fc fragments, the initial rates of disappearance of cytoplasmic κ-chain and Fc fragments were different. This difference was shown by the increase in the κ/Fc ratio during the first hour after treatment with puromycin followed by gradual return to control ratios after 5 hr (insert, Fig. 3).

To ensure that the rapid disappearance of M-Ig was not an artifact specific to puromycin, the studies were repeated with cyclohexamide (50 μg/ml), and similar results were obtained.

To evaluate what percentage of Ig appearing in the culture medium could be caused by membrane turnover, cells were incubated in serum-free medium for 4 hr, and the total Ig in the medium was measured. In 4 hr, 2.25×10^{-7} g of κ chain appeared in the medium for each 1.0×10^6 cells in culture; 1.0×10^6 cells had a total of 2.6×10^{-8} g of κ chain on their surfaces. Using a half-disappearance time of 45 min, 6.9×10^{-8} g of IgG equivalents could appear in the medium in 4 hr and could account for only 3% of the total. The fate of cytoplasmic Ig after treatment of cells with puromycin was determined by

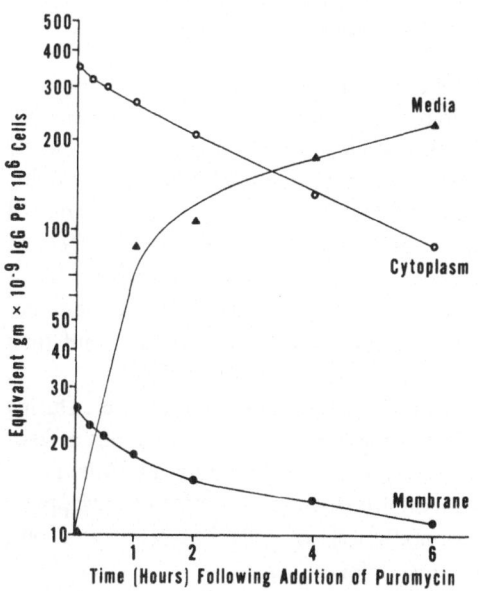

Figure 4. Logarithmically growing cells were washed three times in Earle's salts and suspended in fresh prewarmed (37°C) medium which contained puromycin (50 μg/ml), and the amount of Ig lost from the membrane or cytoplasm and that which appeared in the medium were determined at intervals. o———o cytoplasmic Ig; •———• M-Ig; ▲———▲ Ig in the medium.

simultaneously measuring its accumulation in the medium and disappearance from the cytoplasm. A balance study is shown in Fig. 4. After treatment of cells with puromycin, almost all the Ig lost from the intracellular pool appeared in antigenic form in the medium. The fate of M-Ig, because of the small amounts present, cannot be followed in a similar fashion unless a specific marker which segregates it from total cellular Ig is found. The discovery of this kind of marker would greatly accelerate progress in the study of biological and physical properties of M-Ig.

C. Amount and Half-Disappearance of M-Ig and Cytoplasmic Ig in Synchronized "G$_0$" Cells

To determine if the half-disappearance time for M-Ig differed in growing cells as opposed to cells arrested in the stationary phase, its amount and rate of disappearance were studied. Stationary cells were arrested in the "G$_0$" phase of the cell cycle (see below).

The amount of M-Ig and cytoplasmic Ig in "G$_0$" cells was approximately 70 and 10%, respectively, of that in random cells. The half-disappearance of M-Ig after treatment of cells with puromycin was again 45 min.

The biological significance of the short half-disappearance time for M-Ig is at present unknown but may be related to "modulation" of the immune response.

D. Half-Disappearance of M-Ig and Cytoplasmic Ig in Logarithmically Growing Cells Treated with Actinomycin D

To determine the half-disappearance of M-Ig after treatment of cells with an inhibitor of transcription of RNA, logarithmically growing cells were treated with actinomycin D (5 μg/ml) and membrane and cytoplasmic κ chains and Fc fragments were measured at intervals after drug treatment. Within the first hour, membrane and cytoplasmic κ chains and Fc fragments increased above control values (Fig. 5). This is consistent with our previous report that an early effect of antinomycin D in cultured lymphocytes was to increase the rate of Ig synthesis (Lerner and Hodge, 1971). During the next 7 hr, the amount of Fc fragment and κ chain in the membrane decreased to approximately 20% of control and then declined more slowly for the next 40 hr (Fig. 5) (Lerner et al., 1972).

In contrast to M-Ig, cytoplasmic Ig declined with two distinct rates. During the first 8 hr, cytoplasmic κ chains and Fc fragments increased to about 50% of control, and then the rate changed, and they decreased only an additional 20% over the next 42 hr (Fig. 5). Another possibility for the two different disappearance rates of Ig, that there were two populations of cells, has been ruled out, since the experiments were repeated with cell clones.

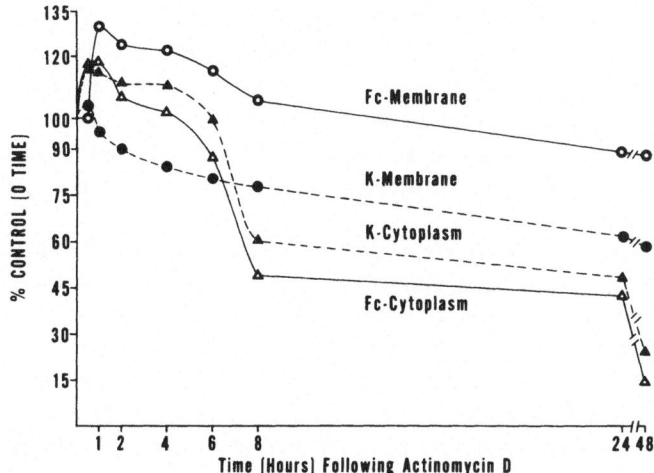

Figure 5. The amount of M-Ig and cytoplasmic Ig was determined for a culture of logarithmically growing cells (0 time), and then actinomycin D was added to a final concentration of 5 µg/ml. At intervals, cell viability and the amount of M-Ig and cytoplasmic Ig were determined. ●−−−−● κ-chain membrane; ○————○ Fc-γ fragment memrane; ▲−−−−▲ κ-chain cytoplasm; △————△ Fc-γ fragment cytoplasm.

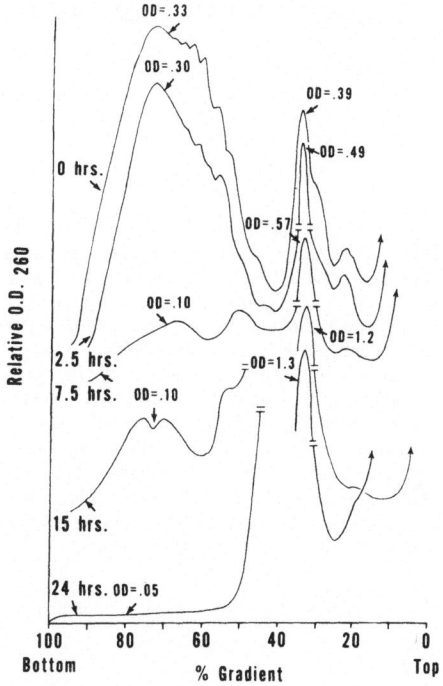

Figure 6. Polyribosomes were prepared from 5.0×10^7 logarithmically growing cells and sedimented through 7.5-45% w/w linear sucrose gradients (0 hr). Actinomycin D (5 µg/ml) was added, and at intervals after addition of the inhibitor polyribosomes were prepared. The absolute OD_{260} and time after addition of actinomycin D for each profile are indicated by arrows.

A likely explanation for these results is that M-Ig and cytoplasmic Ig are controlled separately by the cell. If M-Ig and cytoplasmic Ig have the same primary structure, this might be an example in cell biology in which the same polypeptide would have different fates depending on its cellular control. This kind of control could occur at a translational level and may be an important system for understanding the process of translational regulation of polypeptide synthesis in eukaryotic cells.

Since one explanation for the above results was that the M-RNA which coded for M-Ig was long-lived, comparison of its half-life to total cellular M-RNA in cultures of our lymphocytes was of interest. Accordingly, polyribosome profiles and total nascent polypeptide synthesis were studied after treatment with actinomycin D (Fig. 6). By 24 hr after treatment with actinomycin D, the OD_{260} of polyribosomes had decreased to approximately 15% of control and the amount of 74S monosomes had increased over threefold. The monosome/polysome ratio increased from approximately 1 to 26. During the first 15 hr after treatment with actinomycin D, the polyribosome profile shifted from monophasic to biphasic. A similar shift in polyribosome profile was seen when random cells were synchronized in "G_0" (see below). The amount of nascent polypeptide synthesis decreased with an average half-life of 4 hr to approximately 4% of control in 24 hr (Fig. 7). If the differences in the disappearance of M-Ig and cytoplasmic Ig are due to cessation of synthesis of their respective M-RNAs, these results suggest that, in lymphocytes, the M-RNA which codes for

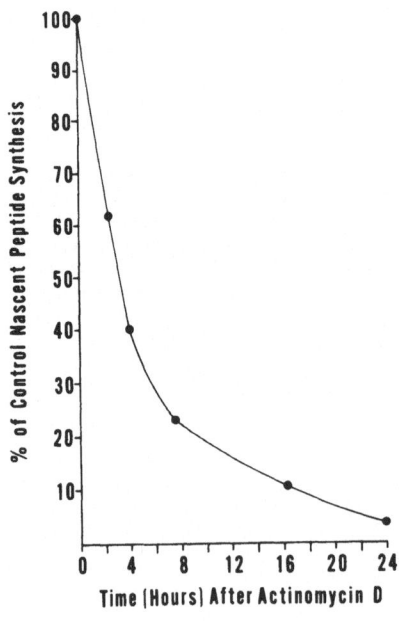

Figure 7. To determine each point on the curve, a culture of logarithmically growing cells was divided into 2. One received actinomycin D (5 μg/ml) for the interval indicated, and the other was untreated; 5.0 × 10^7 cells were pulse-labeled with 100 μCi/ml of leucine -4,5-^{3}H polyribosomes were prepared immediately and sedimented through 7.5-45% sucrose gradients, and fractions were collected (Lerner and Hodge, 1971). Radioactivity associated with the polyribosomes (approximately bottom 40% of the gradient) was determined and expressed as a percentage of nascent peptide synthesis of the untreated culture.

M-Ig is more long-lived than 96% of other cellular M-RNAs. The long-lived nature of the M-Ig M-RNA might offer technical advantages for its isolation. For example, it may be possible to pulse-label cells and let the bulk of cellular M-RNA decay, thus leaving relatively purified Ig M-RNA.

IV. MOLECULAR EVENTS DURING THE REST TO PROLIFERATION TRANSITION IN LYMPHOCYTES

In lymphocytes *in vivo,* the events resulting in cellular proliferation can be categorized conveniently into two fundamental steps: (1) the initial "induction" by immunogen, and (2) the subsequent entry of "G_0" cells into the cell cycle. Since a pure population of lymphocytes capable of induction by specific immunogen is not yet available, the first step cannot be investigated at the molecular level. To investigate the molecular events involved in the second step of the process, it would be necessary to have populations of resting ("G_0") cells which could be manipulated to enter the cell cycle and begin synthesis of heavy- and light-chain polypeptides. The present illustrations from studies in our laboratory demonstrate that a continuous line of cultured diploid lymphocytes arrested in the stationary phase of growth can be used for such an investigation. Specifically, the transition from resting to proliferating cell ("G_0" → G_1; see below) can be defined in terms of the cell cycle, polysome profiles, and the synthesis of specific polypeptides. In addition, data relating the control of total protein synthesis to specific Ig polypeptides can be obtained. These results combined with reports in the literature indicate that this synchronized cell system may be ideal for an investigation of regulation in mammalian cells because specific polypeptides and their corresponding M-RNAs may be isolated in relatively pure form.

A. Synchronization of Cultured Lymphocytes

Phase microscopy revealed that, as cultures of WIL_2 lymphocytes aged, the cells became smaller (Lerner and Hodge, 1971). Using DNA synthesis and viable cell counts as criteria, phases of cell rest (stationary phase) and logarithmic growth in these cultures (log) could be defined (Fig. 8a). After establishment of the culture, the synthesis of DNA was maximum at 2 days, and by the eighth day the rate of DNA synthesis was approximately 2% of this maximum. Viable cell counts increased to a maximum by 6 days and remained constant for approximately 3 days. These 8- to 10-day-old cultures, composed primarily of small lymphocytes not synthesizing DNA, were considered to be a resting population. To define the transition from this resting population to active proliferation in terms of the cell cycle, resting cells were harvested, resuspended

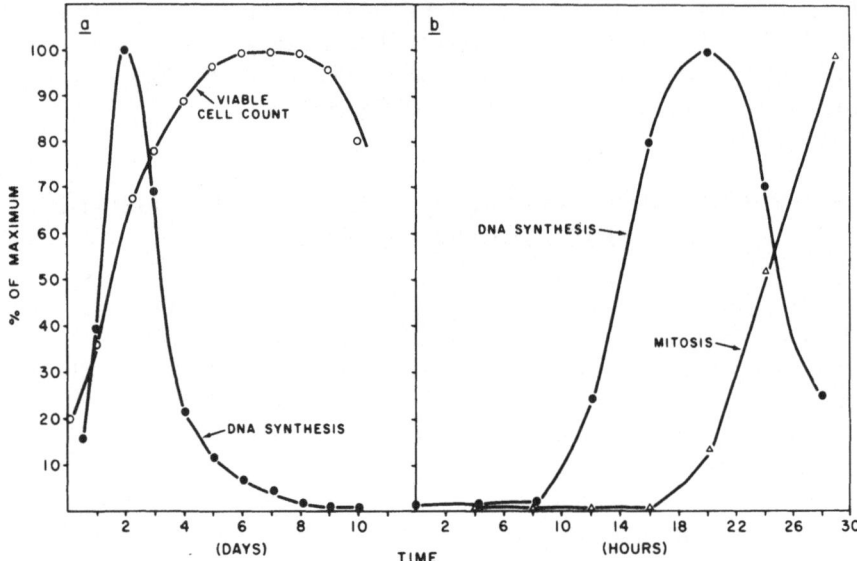

Figure 8. (a) Definition of rest vs. log cultures of WIL$_2$ lymphocytes. Cultures were established at a cell count of 2×10^5/ml. At 24 hr intervals, viable cell count was ascertained and DNA synthesis determined by incubating 2.0 ml aliquots of cells with 2 μCi of thymidine-^{14}C. The data are expressed as the percent of the maximum per 1×10^5 viable cells. (b) Cell cycle analysis. Resting lymphocytes from a 9-day-old culture (see a) were resuspended at 2×10^5 cells/ml in fresh, prewarmed (37°C) Eagle's medium containing 0.2 μg/ml of colchicine. At 1 hr intervals, DNA synthesis was determined as in (a) and the number of cells in mitosis ascertained using phase contrast microscopy. Data are expressed as percent of the maximum rate of DNA synthesis and number of cells in mitosis (Lerner and Hodge, 1971).

in fresh, prewarmed medium, and monitored for DNA synthesis and mitosis (Fig. 8b). There was an 8 hr interval (G$_1$) followed by DNA synthesis (S phase) and, after 18–20 hr, by some cells in mitosis (Lerner and Hodge, 1971). By 28 hr after resuspension, 70–80% of the colchicine-treated cells were arrested in metaphase. In agreement with other cell systems, these data indicated that resting, cultured lymphocytes immediately entered the G$_1$ phase of the cell cycle after resuspension in fresh medium. Of importance was the degree of synchrony in "G$_0$" and the ease with which resting populations could be shifted into proliferation.

B. Phenotypic Expression in Stationary Phase ("G$_0$") and Logarithmically Growing Cells

Changes in cell density and polyribosome profile were considered to represent phenotypic differences between resting and proliferating cells. Because quantitative measurement was possible, density of individual cells was chosen as a parameter. Also, changes in cell density might accompany induction of the immune response (Haskill, 1967; Shortman *et al.,* 1967). Initial studies showed that the optical density profile of the cellular polyribosomes decreased fivefold

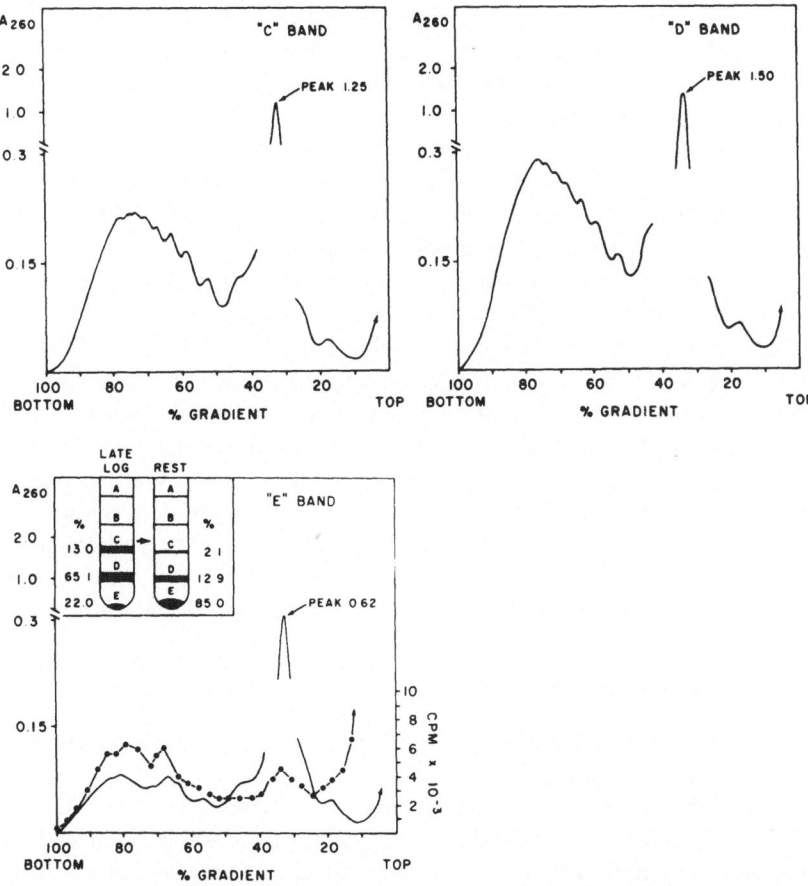

Figure 9. Separation in albumin gradients and polyribosome profiles of rest vs. log cells. From 3- and 10-day-old cultures which had reached viable cell counts of 1.3×10^6/ml (late log) and 3×10^6/ml (rest), respectively, 5×10^8 cells were separated in discontinuous albumin gradients. Three major bands, "C," "D," and "E" (insert), were collected and viable cell count was determined. The data are expressed as the percentage of cells in each of the "C," "D," and "E" bands. From each band separated from a late log culture, 5×10^7 viable cells were resuspended for 15 min in prewarmed (37°C) Eagle's spinner medium containing 1/100 the amount of amino acids and incubated for 6 min with 10 μCi of L-amino acids-^{14}C. Polyribosomes and nascent polypeptides were isolated and separated in 7.5-45% sucrose velocity gradients. Approximately 1.0 ml fractions were collected using a Gilford spectrophotometer with a continuous flow cell, and the TCA-precipitable radioactivity per fraction was determined. —— A_{260}; ●——● c.p.m.

and shifted from a monophasic profile sedimenting at 350S in log cultures to a biphasic profile sedimenting at approximately 180S and 300S in resting cultures. Cells from late log cultures (3 days) and resting cultures (8 days) were separated on the basis of density in discontinuous albumin gradients, and the polyribosome profiles and the synthesis of nascent polypeptides were determined. Cells from these two cultures were separated into three major bands, "C," "D," and "E" (Fig. 9, insert). Of the total cells, the respective bands represented 13.0, 65.1, and 22.0% of late log cultures and 2.1, 12.9, and 85.0% of resting cultures. Thus, there was an approximate fourfold increase in the number of the most dense cells in the resting population.

A comparison of polyribosome profiles from cells at different densities showed that the biphasic profile of sedimentation, characteristic of resting cultures, occurred only in cells from the "E" band, i.e., the most dense (Fig. 9). Polyribosome profiles isolated from cells banding on less dense layers of albumin in the gradient ("C" and "D" bands) resembled the profiles obtained from early log or G_1 cultures (Fig. 9). Since the ratio of maximum optical density of the 74S single ribosomes to the polysomes isolated from the cells in "C," "D," and "E" bands was 5:1, 5:1, and 8:1, respectively, there was some suggestion of an accumulation of free single ribosomes in resting cultures. No evidence for the accumulation of ribosomal subunits was noted in resting cultures. The incorporation of amino acids into nascent polypeptides demonstrated that the polyribosomes isolated from the cells of each band were active in protein synthesis. For simplicity, only the radioactive profile of nascent polypeptide synthesis for "E" band cells is shown (Fig. 9). In cells from both the "C" and "D" bands, a similar correspondence between radioactivity and optical density profile was noted.

C. Polypeptide Synthesis in the "G_0" to G_1 Transition

For the present discussion, we will equate stationary- or rest-phase cells with "G_0" cells (see below).

The relationship of the cellular transition from "G_0" to G_1 to the synthesis of total protein and specific Ig polypeptides was examined. Experiments were initiated with rest-phase cells which, after resuspension in fresh medium, entered the G_1 phase of the cell cycle. The rate of total protein synthesis was maximal by 12 hr and corresponded to a period in S when the rate of DNA synthesis in the culture was approximately 44% of maximum (Fig. 10). By contrast, the maximum rate of total IgG synthesis occurred 8 hr into G_1, when the rate of DNA synthesis was 5% of maximum (Fig. 10a). The specificity of this increase in the rate of total IgG synthesis was suggested by its fivefold increase, whereas total protein increased only twofold. In addition, between 12 and 16 hr after release, the rate of IgG synthesis fell 40%, whereas the synthesis of total protein remained almost constant. Since total cellular IgG polypeptides

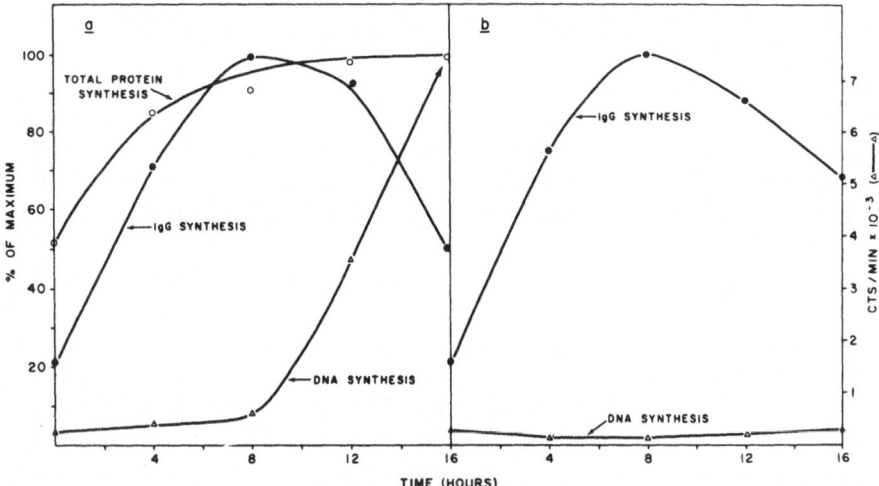

Figure 10. Rate of total and IgG synthesis in the "G_0" to G_1 transition. (a) By resuspension in 500 ml of fresh, prewarmed, complete medium (37°C), 1×10^8 viable cells from a 10-day-old culture were released from rest ("G_0"). At 4 hr intervals, 1×10^7 cells were resuspended in Eagle's spinner medium containing 1/100 the normal amount of amino acids and incubated for 30 min with 10 μCi/ml of L-amino acids-^{14}C, and cytoplasmic polypeptides were prepared. The amount of radioactivity incorporated into cytoplasmic total protein and IgG was determined. DNA synthesis was monitored as in Fig. 8. Data are expressed as percentage of maximum rate of synthesis of total protein, IgG, and DNA. (b) The experiment was carried out exactly as in (a), except that cells were released from rest ("G_0") by resuspension in medium containing 20 μg/ml of cytosine arabinoside. This concentration of inhibitor was also present in the incubation medium containing L-amino acid-^{14}C ○————○ total protein; ●————● IgG; △————△ DNA.

accounted for less than 5–10% of the cytoplasmic protein synthesized in 1 hr, the decrease in the rate of IgG synthesis would not be expected to result in a similar decrease in total protein synthesis.

Synthesis of IgG clearly preceded the onset of DNA synthesis in the transition from resting to proliferating cell. To determine whether the rate of increase or decrease of IgG synthesis differed in the absence of DNA replication, the experiment was repeated in the presence of cytosine arabinoside. Inhibition of at least 98% of DNA synthesis did not alter significantly either the rate of increase or decrease of IgG synthesis (Fig. 10b). These results differed from S-phase cell histone synthesis, which was inhibited within 30 min after the addition of cytosine arabinoside (Borun et al., 1967).

These data and those of others have demonstrated that the rate of synthesis and/or the amount of total Ig present varied during the lymphocyte cell cycle. As the cells entered G_1, the rate of synthesis increased fivefold and then decreased during S and G_2, with the IgG content of the cells the lowest during mitosis (Buell and Fahey, 1969; Takahashi et al., 1969). This was in contrast to M-Ig (see above), which appeared to remain constant, at least in G_1. Our present methods had the advantage that cell synchrony was obtained without the use of inhibitors, and only the rate of synthesis of Ig polypeptides appearing in the cytoplasm was studied. Thus the complexities of assembly and secretion of intact IgG or the fates of synthesized molecules were avoided. *In toto*, these findings concerning the Ig polypeptides support the concept that gene expression during the mammalian cell cycle is a highly ordered sequential process.

Just as there is an order to the synthesis of specific polypeptides during the mammalian cell cycle, periods during which enzyme induction can occur are limited (Martin *et al.*, 1969). In liver cells, tyrosine aminotransferase could be induced by corticosteroid only during the latter two thirds of G_1 and anywhere in the S phase (Martin *et al.*, 1969). Although the present data do not deal with the question of induction of IgG synthesis, it seems reasonable to assume that one might also find a limited period in the lymphocyte cell cycle during which immunogen could induce cellular proliferation. The role of immunogen in this process might be to "induce" resting ("G_0") lymphocytes to pass through a phase in the cell cycle in which the synthesis of heavy- and light-chain polypeptides is obligatory. This obligatory synthesis of Ig polypeptides would be similar to histone synthesis during S, melanin formation in G_1, and increased synthesis of mitotic apparatus proteins in G_2 in other systems (Borun *et al.*, 1967; Whittaker, personal communication; Robbins, personal communication). Indeed, Prescott, in discussing the fact that cells *in vivo* (i.e., neurons and striated muscle) retire from the cell cycle in G_1, has suggested that for most cells control of further proliferation and/or differentiation is mediated by an event during G_1 (Prescott, 1968). An *in vitro* example of this event is the finding by Nilausen and Green (1965) that contact-inhibited mouse fibroblasts were arrested in the G_1 phase of the cell cycle. We would speak of "G_0" rather than G_1 and, for lymphocytes, envision contact with immunogens as the inducing event which initiates entry of a cell into G_1 and subsequent cell proliferation. Recent data of others suggest that a stationary-phase cell and a G_1 cell have physiological differences, and thus the term "G_0" rather than G_1 for cells arrested in the stationary phase of growth may be more appropriate (Becker *et al.*, 1971).

Finally, the present data bear on the question of "recruitment" of antibody-forming cells after induction of an immune response. As cells proceed from "G_0" through G_1, Ig synthesis can increase as much as fivefold without cell division or even DNA synthesis. Thus "recruitment" can be explained as passage of an increasing number of cells to a point in the cell cycle at which the rate of antibody synthesis is sufficient for detection (i.e., during 90 min in a Jerne plaque assay). This contention is supported by the finding that, during induction of an immune response, DNA synthesis and mitosis were not necessary for antibody synthesis by individual cells (Tannenberg, 1967). It also seems clear that some of the observed differences in lymphocyte morphology (i.e., small vs. large or position of sedimentation in an albumin gradient) may simply reflect different phases of the cell cycle.

D. Perturbations Using Actinomycin D and Cyclic AMP

In differentiating between transcriptional and translational control, it would be useful to have agents which rapidly alter the rate of protein synthesis and to correlate this alteration with the rate of synthesis of other macromolecules, e.g., M-RNA of heavy and light chains. Therefore, actinomycin D and cyclic AMP, which have been helpful in probing regulatory mechanisms in other mammalian systems, were examined for their effect on the rate of protein and light-chain synthesis in resting lymphocytes. Actinomycin D at a concentration of 0.1 μg/ml enhanced total protein synthesis 3.5-fold in 120 min (Fig. 11a), and cyclic AMP at a concentration of 1.0 mM caused a fivefold enhancement in a

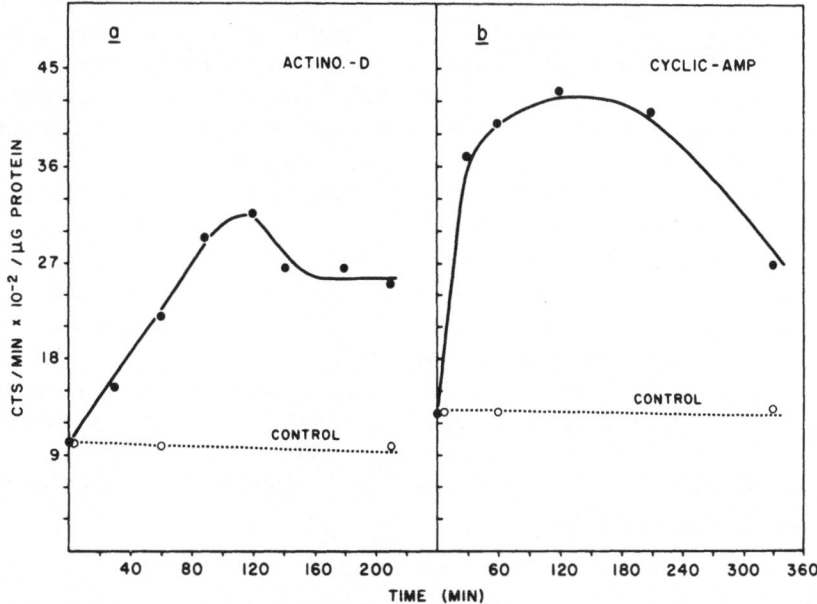

Figure 11. Effect of actinomycin D (a) and cyclic AMP (b) on the rate of protein synthesis in resting ("G_0") lymphocytes. Nine-day-old cultures ("G_0") at a count of 3 × 10⁶ viable cells/ml were treated with either actinomycin D (0.1 μg/ml) or cyclic AMP (1 mM). Untreated cultures served as controls. At 30 min intervals, 5 × 10⁷ cells were resuspended in 1/100 amino acid medium containing 10 μCi/ml of amino acids-¹⁴C incubated at 37°C for 30 min, and cytoplasmic polypeptides were prepared. The amount of TCA-precipitable radioactivity incorporated per microgram of cytoplasmic protein was determined. ●———● c.p.m./μg protein in treated culture; o———o c.p.m./μg protein in control culture.

similar period (Fig. 11b). Although enhancement of protein synthesis was observed with 0.01, 0.1, 2.0, and 5.0 μg/ml of actinomycin D, the maximum effect was with 0.1 μg/ml. The kinetics of the increase of intracellular light-chain synthesis with either agent paralleled that of total protein. In the first 30 min, approximately 85% of the maximum enhancement was noted in the presence of cyclic AMP, whereas only 20% was recorded with actinomycin D. The maximum rate of protein synthesis after treatment with either agent was equal to or greater than synthesis after rest cells were released from "G_0" by placement in fresh medium. However, the maximum rate of protein synthesis was reached in 120 min instead of 12 hr after release from "G_0." Since the kinetics of increased synthesis of total protein and light chain were parallel, those agents did not appear to act specifically in altering the synthesis of Ig protein. Separation of total cytoplasmic polypeptides in acrylamide gels synthesized at 0, 60, 120, and 240 min after treatment with actinomycin D demonstrated that enhanced synthesis of the entire spectrum of cytoplasmic proteins occurred.

The present study demonstrated that both actinomycin D and cyclic AMP were capable of increasing the rate of protein synthesis in resting lymphocytes. In the case of actinomycin D, the effect was general, and the rate of synthesis of a number of polypeptides was increased. Similar results have been reported for other mammalian systems, and it has been postulated that the stimulatory effects of both of these agents operate at the post-transcriptional level (Garren *et al.*, 1964). Although we have not differentiated between transcriptional or post-transcriptional effects, the ability to isolate specific polypeptides (i.e., light chains) and their corresponding M-RNAs from lymphocytes would seem to offer a direct approach to the problem. For example, if the rate of synthesis of light chain increases in the presence of cyclic AMP while no new synthesis of light-chain M-RNA occurs, a strong argument could be made for an effect of this agent at the post-transcriptional level. The demonstration that cyclic AMP increased the rate of protein synthesis in "G_0" lymphocytes was especially interesting in view of the presumed role of this molecule in the control of hormone synthesis and the recent demonstration that phytohemagglutinin caused as much as a threefold increase in cyclic AMP levels in human lymphocytes (Smith *et al.*, 1970, 1971).

V. STUDIES ON THE "LINKAGE" BETWEEN THE PLASMA MEMBRANE AND CELLULAR GENES

Conceptually, the union of immunogen with M-Ig could alter gene expression by several mechanisms. Among these are (1) liberation of a small intermediary molecule such as cyclic AMP which might act at the level of either transcription or translation; (2) initiation of translation of membrane-associated

RNA ultimately resulting in synthesis of a regulatory polypeptide; or (3) initiation of either replication or transcription of membrane-associated DNA leading to gene amplification or cell regulation. The feasibility of the first two mechanisms was illustrated by the role of cyclic AMP in the regulation of hormone synthesis and the association of RNA with cytoplasmic membranes in other systems (Attardi *et al.*, 1969; Glick and Warren, 1969). Although all these possibilities are being investigated in this laboratory, the present discussion reviews our initial attempts to determine if there is an association which might support the third mechanism between the plasma membrane and DNA. The WIL$_2$ line was studied (Lerner *et al.*, 1971*a*). A species of cytoplasmic DNA apparently associated with the plasma membrane was described (Lerner *et al.*, 1971*b*). This DNA differs from the cytoplasmic DNA studied by others by location in the cell and characteristics which distinguish it from nuclear and

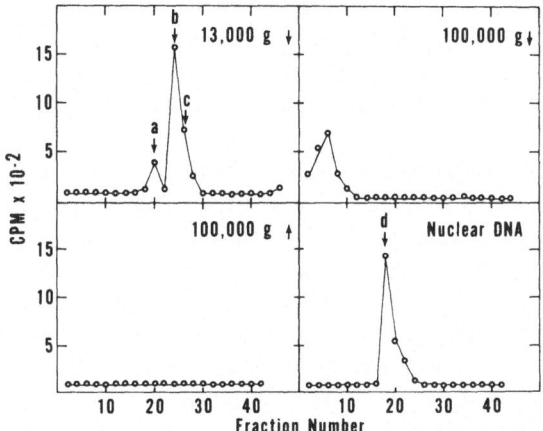

Figure 12. Equilibrium centrifugation of DNA from cell fractions in EtBr-CsCl density gradients. DNA from 5.0×10^8 cells was labeled with ^{14}C for two cell generations as in Table IV. Cells were collected by centrifugation for 2 min at $800 \times g$, washed three times in cold (4°C) Earle's salts, suspended in 2.0 ml of RSB buffer (1.0×10^{-2}M NaCl, 1.5×10^{-3}M MgCl$_2$, 1.0×10^{-2}M tris-HCl, pH 7.4), and lysed by the addition of 0.2 ml of 5% NP-40. Nuclei were sedimented from the lysate by centrifugation at $1000 \times g$ for 15 min at 4°C. The supernatant fluid was carefully decanted, and material sedimenting at $13,000 \times g$ ($13,000\ g \downarrow$) was prepared by centrifugation at 4°C for 15 min. Material sedimenting at $100,000 \times g$ ($100,000\ g \downarrow$) was prepared from the $13,000 \times g$ supernatant fluid by centrifugation for 60 min at 4°C. The final supernatant fluid was decanted ($100,000\ g \uparrow$). Nuclei and the $13,000\ g \downarrow$ and $100,000\ g \downarrow$ fractions were suspended in 1.5 ml of STE buffer (1.0×10^{-1} M NaCl, 1.0×10^{-3} M EDTA, 5.0×10^{-2} M tris-HCl, pH 7.4) and made 1% in sodium dodecyl sulfate. Each cell fraction was selectively extracted. Supernatant fluids were centrifuged in 3.0 ml of CsCl (ρ=1.571) containing 100 μg/ml of EtBr for 48 hr at 31,000 rev/min in the SW50L rotor. Fractions were collected, and the radioactivity in a 0.02 ml sample of each fraction was measured after precipitation with 5% Cl$_3$CCOOH. a, b, c, and d (arrows) indicate fractions that were studied by velocity sedimentation (see Fig. 14).

mitochondrial DNA (Bell, 1969; Bond *et al.,* 1969; Clayton and Vinograd, 1967; Fromson and Nemer, 1960; Hudson and Vinograd, 1967; Müller *et al.,* 1970; Schneider and Kuff, 1969; Williamson, 1970).

A. Nature of Cytoplasmic DNA

To determine if nonmitochondrial cytoplasmic DNA existed in human lymphocytes and to investigate its subcellular localization, cellular DNA was uniformly labeled with thymidine-2-^{14}C, and cells were lysed with the nonionic detergent NP-40 (Lerner and Hodge, 1971). This detergent lysed the plasma membrane but not the nuclear membrane, as shown by phase contrast microscopy, and nuclei could be quantitatively removed from lysates by low-speed centrifugation. The lysate obtained after removal of nuclei was separated into several fractions. To ensure removal of any contaminating, high molecular weight DNA, material obtained from the individual fractions was selectively extracted utilizing 1% SDS followed by precipitation with 1 M NaCl (Hirt, 1967). Linear and supercoiled DNA molecules were then separated in ethidium bromide-cesium chloride (EtBr-CsCl) equilibrium density gradients (Radloff *et al.,* 1967).

Virtually all cytoplasmic DNA was found in a fraction sedimenting at 13,000 x *g,* and of this less than 10% was supercoiled (Fig. 12). Cytoplasmic DNA represented approximately 0.5% of the total cellular DNA which incorporated isotope during two cell generations. A small amount of DNA was also found in the fraction sedimenting at 100,000 x *g* (Fig. 12), but its unusually high density in EtBr-CsCl suggested association with more dense molecules such as RNA. The nature of this latter DNA is presently being investigated and so is not considered in this report. DNA was not found in the supernatant fluid from the 100,000 x *g* fraction (Fig. 12).

DNA from peak fractions (Fig. 12) of the EtBr-CsCl gradients was further characterized by velocity sedimentation in CsCl (Fig. 13). Two species of DNA which sedimented at approximately 35S and 16S were recovered from EtBr-CsCl gradients of cytoplasmic fractions (Fig. 13). The 35S DNA had the expected sedimentation characteristics of mitochondrial DNA, whereas the 16S DNA presumably represented a unique species. No 16S component was found in DNA that remained after selective extraction of nuclei (Fig. 13).

B. Association of DNA with Cytoplasmic Membranes

Since the bulk of cytoplasmic DNA sedimented at 13,000 x *g,* its association with larger cytoplasmic structures was suggested. To study the nature of the complex, cellular DNA was uniformly labeled with thymidine-2-^{14}C and cells were lysed with NP-40. Aliquots of cytoplasm were either not treated further or

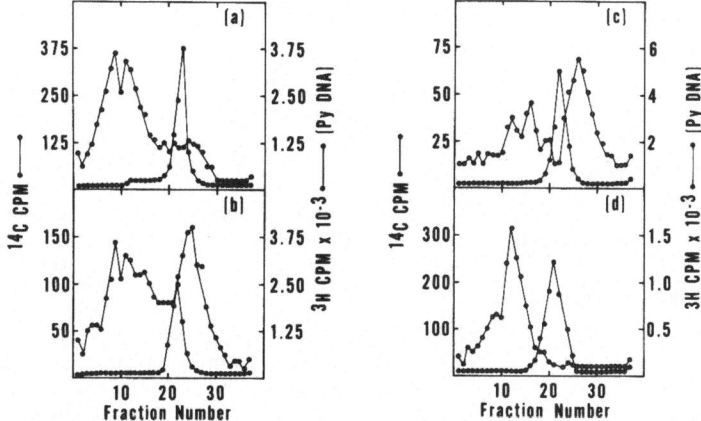

Figure 13. Velocity sedimentation of DNA from cell fractions. Fractions 19-21 (a), 23-25 (b), and 26-29 (c) from the 13,000 g ↓ DNA and fractions 18-24 (d) from nuclear DNA were collected separately from the EtBr gradients shown in Fig. 12 and dialyzed for 24 hr against 0.001M EDTA plus 0.02 M tris-HCl (pH 8.0). A 0.2 ml sample of the DNA in each pool was sedimented through a CsCl plus 0.01 M EDTA (pH 7.5) solution ($\rho = 1.50$ g-cm^{-3}) for 2.5 hr at 35,000 rev/min at 25°C in the SW50.1 rotor. o———o ^{14}C; •———• ^{3}H-labeled polyoma (Py) DNA marker (20S).

digested with phospholipase C or sodium deoxycholate (DOC) and sedimented in linear sucrose gradients. All untreated DNA sedimented to the lower one third of the gradients between densities of 1.198 and 1.135 g-cm^{-3}, whereas treatment with DOC caused release of 90% of the rapidly sedimenting DNA (Fig. 14). Treatment with phospholipase C caused the DNA to sediment slightly faster and sharpened the band of sedimentation (Fig. 14). The position of sedimentation of the DNA membrane complex varied depending on the particular lot or exact concentration of NP-40 relative to the number of cells. The optimum concentration of NP-40 for preservation of the rapidly sedimenting DNA membrane complex varied between 0.1 and 0.5% for 5.0 x 10^8 cells suspended in 2.0 ml of RSB.

The fact that each cell has Ig associated with its plasma membrane (Lerner *et al.*, 1971*a*) was utilized to locate the position of sedimentation of plasma membrane fragments in the sucrose gradients described above. Cells were incubated for 60 min at 4°C with 1 μg of immunospecifically purified sheep anti-human IgG labeled with ^{125}I. After the sheep antibody had bound to the IgG on the plasma membrane, cells were lysed and the cytoplasmic fraction was sedimented in sucrose gradients as described. The ^{125}I-labeled sheep antibody which was bound to plasma membrane fragments sedimented to the same position in the gradients as the DNA (Fig. 14).

The above findings suggested that DNA was bound to the plasma mem-

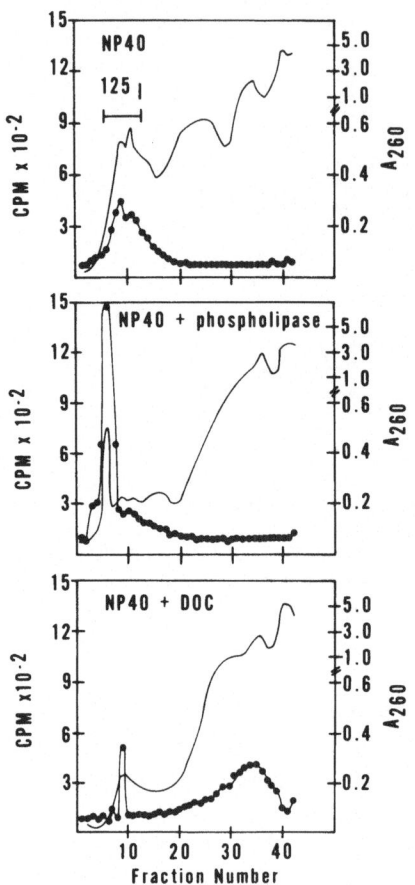

Figure 14. Velocity sedimentation of the DNA membrane complex. The DNA from 6.0×10^8 cells was uniformly labeled with ^{14}C for two cell generations as in Table IV. Cells were collected and lysed by NP-40, and nuclei were removed as described in Fig. 12. Aliquots of the supernatant fluid were either not treated further, made 1% with deoxycholate, or treated with phospholipase C (200 μg/ml) at 37°C for 30 min. In a parallel experiment, 6.0×10^8 cells were washed three times in Earle's salts at 4°C and resuspended in 50 ml of Earle's salts containing 2.0×10^{-6} g of immunospecifically purified sheep anti-humen IgG labeled with ^{125}I (1.0×10^6 c.p.m./1.0×10^{-6} g protein). Cells and antibody were incubated at 4°C for 60 min, sedimented at 800 × g for 2 min, and washed three times in cold (4°C) Earle's salts. Cells were lysed and nuclei removed as described above. Lysates were sedimented at 95,400 × g for 45 min through a 36 ml 7.5-45% (w/w) linear sucrose gradient in the SW27 rotor at 4°C. Fractions were collected through a Gilford recording spectrophotometer (0.5 cm light path), and radioactivity was determined. ——— A_{260}; •———• c.p.m. ^{14}C |——| position of sedimentation of ^{125}I.

brane. To confirm this, DNA was uniformly labeled with thymidine-2-^{14}C and plasma membranes were prepared by both the fluorescein mercuric acetate (FMA) and zinc chloride (ZnCl) methods (Warren *et al.*, 1966). These preparations contained predominately intact or fragmented plasma membranes without morphological evidence for contamination with other subcellular structures. Plasma membranes prepared by both methods were associated with TCA-precipitable radioactivity. The nature of the DNA membrane association was investigated by studying some of the factors which could release radioactivity from purified plasma membranes. The results listed in Table III illustrate that DNA could be quantitatively released by DOC and pronase and suggest that it was on the surface of the membrane, since it was completely susceptible to deoxyribonuclease. The fact that the DNA could be released from the membranes by pronase but not by phospholipase C (Fig. 14) suggests a protein-DNA linkage in the membrane.

Table III. Release of Plasma Membrane-Associated DNA

| | Percent c.p.m. released from membrane | |
Treatment	Total released	Released TCA precipitable
DOC	96	100
Pronase	99	100
DNAse	99	2
RNAse	< 2	–

The DNA released from purified plasma membranes (ZnCl method) or membrane fragments obtained after lysis of cells with NP-40 was compared and characterized with respect to sedimentation equilibrium in EtBr-CsCl and neutral and alkaline CsCl and with respect to sedimentation velocity. No supercoiled DNA was detected. DNA released from membrane fragments had a buoyant density in CsCl equal to that of nuclear DNA (1.699 g cm^{-3}) (Fig. 15). The

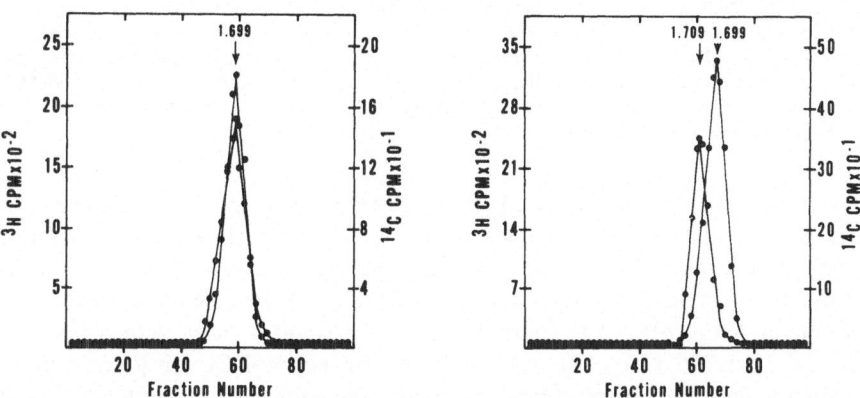

Figure 15. Equilibrium centrifugation of membrane DNA and nuclear DNA. Plasma membrane fragments were prepared from a NP-40 lysate of 7.0×10^8 cells which had been labeled for two cell generations with thymidine-2-^{14}C (0.1 μCi/ml, 50 mCi/mM). The membranes were made 1% with SDS, and the DNA was extracted by the Hirt procedure. To prepare nuclear DNA, random cells were labeled for 6 hr with thymidine-methyl-^{3}H (2.0 μCi/ml, 15 Ci/mM). Purified nuclei were obtained after lysis of cells with NP-40, and the DNA was extracted four times with phenol saturated with 0.01 M EDTA plus 0.2 M tris-HCl (pH 8.0). Left: A sample of ^{3}H-labeled nuclear DNA and ^{14}C-labeled DNA released from membrane (M-DNA) was mixed with solid CsCl and adjusted with 0.02 M tris-HCl (pH 8.0) to a density of approximately 1.70 g-cm^{-3} and to a volume of 3.0 ml. The solution was certrifugated for 60 hr at 31,000 rev/min at 25°C in the SW50L rotor. o————o ^{3}H (nuclear DNA); •————• ^{14}C (M-DNA). Right: A sample of ^{3}H-labeled nuclear DNA and ^{14}C-labeled polyoma DNA was mixed with solid CsCl, adjusted to 1.71 g-cm^{-3}, and centrifuged as above. o————o ^{3}H DNA; •————• ^{14}C [polyoma (Py) DNA marker, ρ = 1.709].

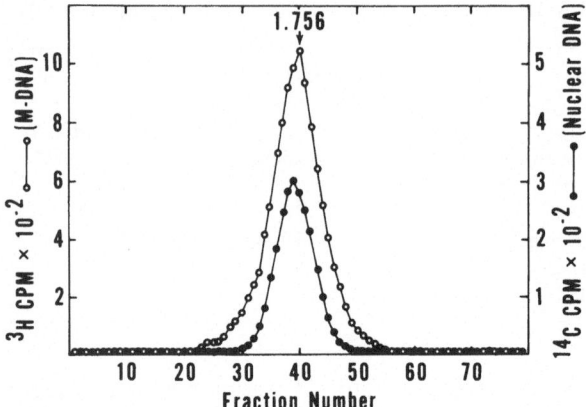

Figure 16. Equilibrium centrifugation of membrane DNA and nuclear DNA in alkaline CsCl. Plasma membrane fragments were prepared from a NP-40 lysate of 7×10^8 cells which had been labeled for 6 hr with thymidine-methyl-^{3}H (2.0 μCi/ml, 15 Ci/mM) and separated in sucrose gradients as described. The membranes were made 1% with SDS, and the DNA was extracted twice with phenol as described in Fig. 15. To prepare nuclear DNA, random cells were labeled for two cell generations with thymidine-2-^{14}C (0.1 μCi/ml, 50 mCi/mM). Purified nuclei were obtained after lysis of cells with NP-40 and made 1% with SDS, and the DNA was extracted four times with phenol. A sample of ^{3}H-labeled membrane DNA (M-DNA) and ^{14}C-labeled nuclear DNA was mixed with solid CsCl, adjusted with 0.5 M Na$_3$PO$_4$ to pH 12.4, and with distilled water to a density of 1.750 g-cm^{-3} and a volume of 3.2 ml. The solution was centrifuged for 48 hr at 35,000 rev/min (25°C) in the SW50L rotor. Fractions were collected, and TCA-precipitable radioactivity was measured. o———o ^{3}H (M-DNA); •———• ^{14}C (nuclear DNA).

density for membrane-associated DNA corresponded to a mole fraction of guanine plus cytosine of 40%, which is in good agreement with the results of other authors who have measured human spleen cell DNA (Schildkraut *et al.,* 1962). When nuclear and membrane-associated DNA were centrifuged to equilibrium in CsCl (pH 12.4), both banded as a single species with a buoyant density of 1.756 (Fig. 16). DNA released by either DOC or pronase from purified membranes or membrane fragments sedimented with a mean coefficient of 16S (Fig. 17). This would correspond to a molecular weight for a linear DNA molecule of about 3.0×10^6 daltons. By contrast, nuclear DNA deliberately sheared by passage ten times through a 22-gauge needle sedimented as a band at 23S.

C. Synthesis of Membrane-Associated DNA in Synchronized Cells

To compare the synthesis of membrane-associated DNA to that of nuclear DNA in different phases of the cell cycle, cells were uniformly labeled with thymidine-2-^{14}C and then synchronized in the "G_0" phase of the cell cycle (Lerner and Hodge, 1971). After release from "G_0" arrest, cells were pulse-labeled with thymidine-^{3}H during the G_1 and S phases of the cell cycle, and membrane fragments and nuclei were prepared after lysis of cells with NP-40. The membrane fragments and nuclear DNA were selectively extracted, and the DNA was sedimented to equilibrium in EtBr–CsCl gradients in order to remove any possible contamination from supercoiled mitochondrial DNA. The rate of thymidine-^{3}H/thymidine-2-^{14}C incorporated into nuclear and membrane DNA in the G_1 and S phases of the cell cycle was calculated from the peaks obtained in EtBr–CsCl gradients (Table IV). These results indicate that the membrane-associated DNA studied replicated in both G_1 and S, and the ^{3}H/^{14}C ratio for

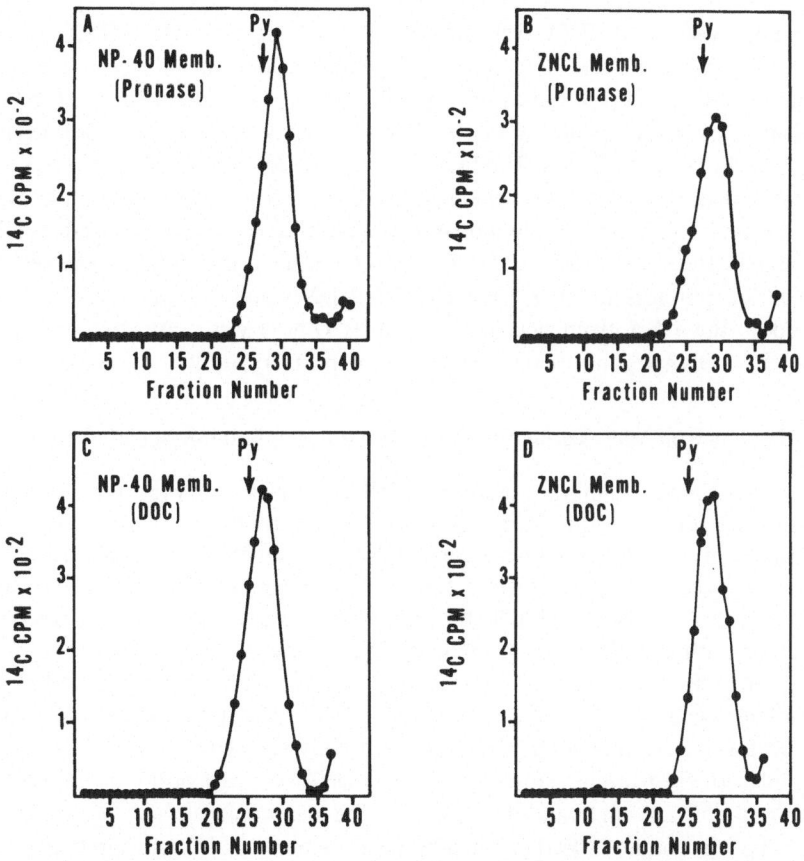

Figure 17. Sedimentation velocity of membrane-associated DNA. Purified plasma membranes (ZnCl$_2$) or membrane fragments from NP-40 lysates were prepared. Aliquots were treated with either deoxycholate (1%) or pronase (500 μg/ml) for 60 min at 37°C; 0.2 ml was layered onto 3.0 ml CsCl (1.50 g-cm^{-3}) and centrifuged for 95 min at 114,000 × g at 25°C. Fractions were collected, and the acid-precipitable radioactivity was determined. Purified polyoma (Py) DNA (20S) was used as a marker.

Table IV. Specific Activity of Plasma Membrane-Associated
DNA During the Cell Cycle

Source of DNA	Phase of cell cycle	
	G_1 ($^3H/^{14}C$)	S ($^3H/^{14}C$)
Nuclei	5	40
Membranes	16	25

membrane-associated DNA was different from that of nuclear DNA in both phases of the cell cycle. If a linear rate of incorporation of thymidine-H^3 is assumed and the incorporation during only 1 hr is considered, then the rate of nuclear DNA synthesis during G_1 is 3% of that in S, whereas cytoplasmic DNA synthesis in G_1 is 16% of S. The fact that nuclear DNA synthesis in G_1 was only 3% of that in S demonstrated the high degree of cell synchrony obtained. These studies seem to exclude the possibility that the membrane-associated DNA originated from the bulk of nuclear DNA. Similar differences between the replication patterns of membrane-associated DNA and nuclear DNA were obtained if the DNA from plasma membranes (ZnCl) was compared to nuclear DNA, or if DNA was not selectively extracted prior to study.

D. Electron Microscopic Studies of Plasma Membrane-Associated DNA

To visualize membrane-associated DNA, samples were prepared by a modification of the spreading procedure of Kleinschmidt and Zahn (1959) and examined in the electron microscope. DNA molecules were clearly associated with membranes (Fig. 18). While some DNA molecules appeared to dissociate from membranes suspended in 1.0 M ammonium acetate, little or no dissociation occurred with membrane suspended in RSB. Every membrane fragment of the approximately 1000 examined was associated with DNA. When purified plasma membranes in RSB were treated for 30 min with deoxyribonuclease (50 μg/ml), no DNA was seen (Fig. 18). DNA was not usually visible unless membranes were treated with formamide. This suggests that DNA was part of the membrane and not simply "stuck" in a random fashion to the membrane surface.

To determine whether membrane-associated DNA was predominately linear or circular, DNA released by SDS treatment of membranes (ZnCl) was examined. More than 99% of the molecules seen were linear. Occasionally, "mini"-circles similar to those described by Radloff et al. (1967) were observed.

Evidence for the association of DNA with the plasma membrane of diploid human lymphocytes has now been obtained in our laboratory. Since this DNA represented about 0.5% of total cellular DNA, a major problem was to demonstrate that contamination of plasma membrane preparations with nuclear, mitochondrial, mycoplasmal, or viral DNA did not account for its presence.

The WIL_2 cell line was in many ways suitable for analysis of plasma membrane-associated DNA. These cells had very little endoplasmic reticulum and few lysosomal structures (Lerner, unpublished observations). Thus con-

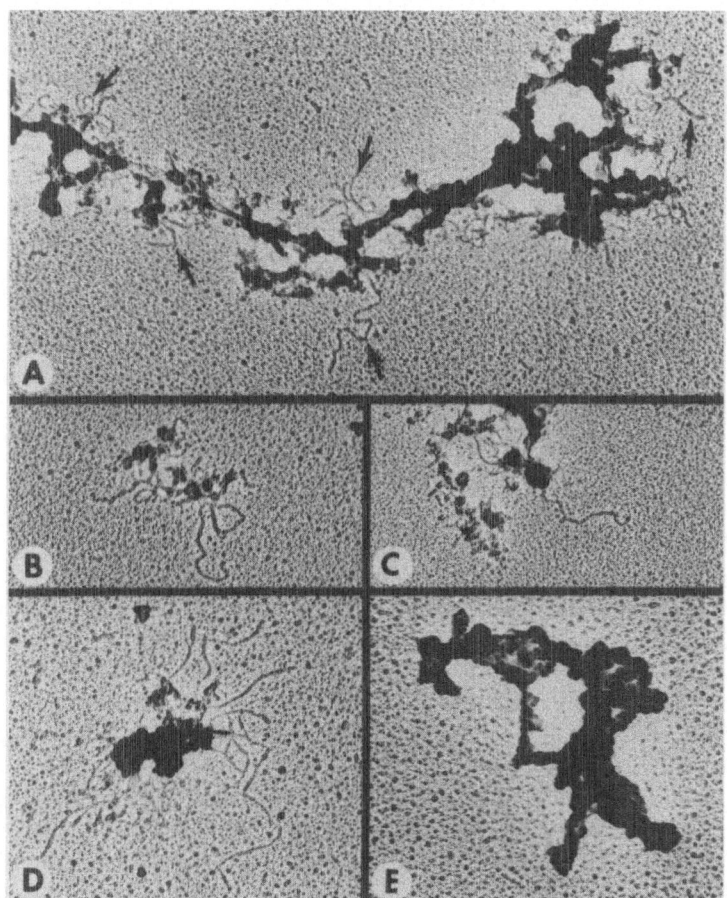

Figure 18. Electron micrographs of plasma membranes. Samples for electron microscopy were suspended in either RSB (see Fig. 12) (A,B,C,E) or 1.0 M ammonium acetate (pH 5.5) (D) and made 42% with formamide and 0.01% with cytochrome c. The solutions were spread on a hypophase of 0.3 M ammonium acetate (pH 5.5), and the films were allowed to age for 10-15 min. The films were transferred to carbon-coated collodion support films on 200-mesh copper grids and rotary-shadowed (120 rev/min) with platinum. DNA molecules are seen associated with membrane fragments (arrows in A, and B-D. × 14,800. No DNA molecules are seen when membranes are treated with deoxyribonuclease (E). (Reduced for reproduction 20%.)

tamination with extraneous membraneous structures or lysosomal enzymes (e.g., nucleases) was minimized.

Contamination from nuclear DNA, while not ruled out, is unlikely. The use of NP-40 to lyse cells, sucrose gradient centrifugation to remove any nuclei not sedimented at 1000 × g, and selective extraction to remove any high molecular weight DNA is a combination of methods which efficiently removes free nuclear DNA. This fact was illustrated by our failure to detect any DNA in the 100,000 × g supernatant fluid. Furthermore, neither deliberate shear nor selective extraction of nuclear DNA from these cells yielded DNA as small as the species associated with plasma membranes. The specific activities of membrane-bound and nuclear DNA and the different times of synthesis during the cell cycle would exclude any random "leak" of nuclear DNA.

The probability that we have isolated a membrane-associated DNA replication complex from nuclei is unlikely because, relative to nuclear DNA, more membrane-associated DNA was synthesized in G_1 than in S. Also, in S the specific activity of pulse-labeled membrane-associated DNA was less than that of total nuclear DNA, and this is the opposite of what would be expected of DNA from a replication complex. The possibility that a unique species of DNA was synthesized in the nucleus and exited to associate with the plasma membrane cannot be excluded, but this would be of considerable interest.

Contamination from mitochondrial DNA seems unlikely since few circular and no supercoiled molecules were detected in DNA released from membranes. Furthermore, the buoyant density of our membrane-associated DNA in neutral and alkaline CsCl was different from that of mitochondrial DNA. The density of human mitochondrial DNA in neutral CsCl differed from that of nuclear DNA by 0.007 g-cm^{-3}, whereas in alkaline CsCl the complementary strands of mitochondrial DNA separated into "heavy" and "light" components with respective densities of 1.766 and 1.727 g-cm^{-3} (Corneo et al., 1968). By contrast, the membrane-associated DNA described here had a density in neutral CsCl equal to that of nuclear DNA (1.699 g-cm^{-3}), and in alkaline CsCl only a single component was observed.

Contamination from mycoplasmal DNA can be excluded. These cells were shown repeatedly by electron microscopy to be free from structures which resembled mycoplasma (Lerner and Feldman, unpublished observations). Mycoplasma was not cultured from these cells in attempts to do so by Dr. Mary Pollock of this institution. Furthermore, the guanine plus cytosine content and size of this DNA were quite different from those of almost all mycoplasma (Hayflick and Stanbridge, 1967; McGee et al., 1967; Williams et al., 1969), particularly those known to contaminate cell cultures.

The possibility that these cells carried a latent viral genome is more difficult to exclude. These cells were repeatedly studied by electron microscopy and serological methods, and no evidence for morphological structures resembling virus or production of viral antigens was obtained (Lerner, unpublished observations). Also, any viral genome carried would have to have the same guanine plus cytosine content as host nuclear DNA and be distributed throughout the plasma membrane.

Finally, the fact that 100% of the membrane fragments obtained from purified plasma membrane preparations could be shown to be associated with DNA made it unlikely that all the DNA studied originated from other contaminating DNA-membrane complexes. For example, if we argue that our preparations contained 10% mitochondrial DNA complexes, then 90% of membranes should have been free from DNA. A similar argument can be made against impurities from DNA-membrane complexes of nuclear or mycoplasmal origin.

Although the function or precise origin of this plasma membrane-associated DNA is unknown, several speculations in relation to immunogen recognition can be considered. If an analogy to the replicon model of Jacob (Jacob, 1966; Jacob et al., 1963) is made, then union between the Ig receptor in the plasma membrane and immunogen might cause a conformational change in the membrane, allowing the DNA to either replicate or be transcribed. The result of replication could be gene amplification, which may accompany lymphocyte differentiation (Krueger and McCarthy, 1970), whereas initiation of transcription might result ultimately in the synthesis of a regulatory polypeptide. Another possibility is that plasma membrane-associated DNA is involved in the transfer of genetic information from cell to cell (e.g., thymus to bone marrow), and in this sense the transfer would be similar to colicinogenic and F factors in bacteria (Nomura, 1967).

An interesting relationship to consider is the ratio of membrane-associated DNA to Ig. The cell line studied has 1.8×10^4 M-Ig molecules per cell (Lerner et al., 1971a). If one assumes 1.0×10^{-11} g of DNA per diploid cell, 0.5% of which is associated with the plasma membrane, then there is approximately one molecule of DNA of 3.0×10^6 daltons for each molecule of M-Ig. DNA of 3.0×10^6 daltons is sufficient to code for the synthesis of about 1700 amino acids. This is larger than necessary for the synthesis of the Ig molecule.

Further studies will be necessary to determine whether the plasma membrane-associated DNA occurs in other cell types or has any role in the immune response. A survey of other cells including tumor cells is in progress and will be reported elsewhere.

VI. SUMMARY AND PROSPECTS FOR THE FUTURE

As more has been learned about the complexity of subcellular organelles in eukaryotics, it has become increasingly more difficult to consider the cell a simple sphere in which macromolecular interactions are under simple diffusion control so that molecules such as messengers, inducers, repressors, and even nucleotides and amino acids randomly find their "home." If this difficulty is real, then the question put squarely in our laps is how are cells compartmentalized so that events at the cell surface influence the morphological nucleus, which is at angstrom distance.

In the present report, our initial efforts to approach this problem are presented. We have begun with a "functional" approach to the behavior of the presumed antigen receptor and have begun a study of the molecule(s) which "inform" the genetic material of a cell what is going on in the environment. These studies are just the beginning, and much more work will be needed for definitive answers. Nevertheless, it is already clear that diploid, continuously growing, suspended cell lymphocytes are an important tool for biochemical and immunological studies of events at cell surfaces.

Other uses for these cells are becoming evident. Since the rate of success in obtaining continuous diploid cultures from any individual is high, cell lines with important genetic markers can be obtained. For example, lines from patients with the Lesch-Nyan syndrome have already been established.

Finally, it has been shown that lines of diploid cells produce immunological mediators such as migration inhibition factor and lymphotoxin (Tubergen *et al.*, 1971). It should be possible to obtain and purify large amounts of such substances from these cultures. If these mediators act at the surface of cells, then they will become important additional reagents in the investigation of how events at cell surfaces alter gene expression.

ACKNOWLEDGMENT

This is publication No. 533 from the Department of Experimental Pathology, Scripps Clinic and Research Foundation, La Jolla, California. This research was supported by a grant from the Council for Tobacco Research, USPHS Grant AI-07007, and USPHS Grant CA-10596.

REFERENCES

Attardi, B., Cavioto, B., and Attardi, G. (1969). *J. Mol. Biol.* **44**:47.
Becker, H., Stanners, C. P., and Kudlow, J. E. (1971). *J. Cell. Physiol.* **77**:43.
Bell, E. (1969). *Nature* **224**:326.
Bond, H. E., Cooper, J. A., Courington, D. P., and Wood, J. S. (1969). *Science* **165**:705.
Borun, T. W., Scharff, M. D., and Robbins, E. (1967). *Proc. Natl. Acad. Sci.* **58**:1977.
Buell, D. N., and Fahey, J. L. (1969). *Science* **164**:1524.
Byrt, P., and Ada, G. L. (1969). *Immunology* **17**:503.
Cerottini, J. -C. (1968). *J. Immunol.* **101**:433.
Choi, K. W., and Bloom, A. D. (1970). *Science* **170**:89.
Clayton, D. A., and Vinograd, J. (1967). *Nature* **216**:652.

Corneo, G., Zardi, L., and Polli, E. (1968). *J. Mol. Biol.* **36**:419.
Fromson, D., and Nemer, M. (1970). *Science* **168**:266.
Garren, L. D., Howell, R. R., Tomkins, G. M., and Crocco, R. M. (1964). *Proc. Natl. Acad. Sci.* **52**:1121.
Glick, M. C., and Warren, L. (1969). *Proc. Natl. Acad. Sci.* **63**:563.
Haskill, J. S. (1967). *Nature* **126**:1229.
Hayflick, L., and Stanbridge, E. (1967). *Ann. N.Y. Acad. Sci.* **143**:608.
Hirt, B. (1967). *J. Mol. Biol.* **26**:365.
Hudson, B., and Vinograd, J. (1967). *Nature* **216**:647.
Jacob, F. (1966). *Science* **152**:1470.
Jacob, F., Brenner, S., and Cuzin, F. (1963). *Cold Spring Harbor Symp. Quant. Biol.* **28**:329
Kleinschmidt, A. K., and Zahn, R. K. (1959). *Z. Naturforsch.* **14B**:770.
Krueger, R. G., and McCarthy, B. J. (1970). *Biochem. Biophys. Res. Commun.* **41**:944.
Lerner, R. A., and Hodge, L. D. (1971). *J. Cell. Physiol.* **77**:265.
Lerner, R. A., McConakey, P., Jansen, I., and Dixon, F. J. (1972). *J. Experi. Med.* (in press).
Lerner, R. A., McConahey, P. J., and Dixon, F. J. (1971*a*). *Science* **173**:60.
Lerner, R. A., Meinke, W., and Goldstein, D. A. (1971*b*). *Proc. Natl. Acad. Sci.* **68**:1212.
Martin, D., Tomkins, G. M., and Granner, D. (1969). *Proc. Natl. Acad. Sci.* **62**:248.
McGee, Z. A., Rogul, M., and Wittler, R. G. (1967). *Ann. N.Y. Acad. Sci.* **143**:21.
Müller, W. E. G., Zahn, R. K., and Beyer, R. (1970). *Nature* **227**:1211.
Nilausen, K., and Green, H. (1965). *Exptl. Cell. Res.* **40**:166.
Nomura, M. (1967). *Ann. Rev. Microbiol.* **21**:257.
Prescott, D. M. (1968). *Cancer Res.* **28**:1815.
Radloff, R., Bauer, W., and Vinograd, J. (1967). *Proc. Natl. Acad. Sci.* **57**:1514.
Schildkraut, C. L., Marmur, J., and Doty, P. (1962). *J. Mol. Biol.* **4**:430.
Schneider, W. C., and Kuff, E. L. (1969). *J. Biol. Chem.* **244**:4843.
Shortman, K., Haskill, J. S., Szenberg, A., and Legge, D. G. (1967). *Nature* **216**:1227.
Smith, J. W., Steiner, A. L., and Parker, C. W. (1970). *Fed. Proc.* **29**:369.
Smith, J. W., Steiner, A. L., Newberry, W. M., Jr., and Parker, C. W. (1971). *J. Clin. Invest.* **50**:432.
Takahashi, M., Yagi, Y., Moore, G. E., and Pressman, D. (1969). *J. Immunol.* **103**:834.
Tannenberg, W. J. K. (1967). *Nature* **214**:293.
Tubergen, D., Lerner, R. A., and Feldman, J. D. (1971). *J. Experi. Med.* (in press).
Warren, L., Glick, M. C., and Noss, M. K. (1966). *J. Cell. Physiol.* **68**:269.
Williams, C. O., Wittler, R. G., and Burris, C. (1969). *J. Bacteriol.* **99**:341.
Williamson, R. (1970). *J. Mol. Biol.* **51**:157.

The Antigen-Binding Sites of Immunoglobulins

R. R. Porter

Department of Biochemistry
University of Oxford
Oxford, England

I. INTRODUCTION

Although an antibody molecule possesses many biological properties essential to its role in immune reactions, the one on which all others depend is its capacity to bind specifically with the antigen, the injection of which has stimulated its synthesis. Increasing knowledge of the structure of immunoglobulins—the group of proteins to which all antibodies belong—has made it seem even more remarkable that molecules of identical basic structure should be able to combine specifically with such a wide range of different substances.

The introduction by Landsteiner more than 50 years ago of the technique of raising antibodies to small molecules by coupling them to proteins before injection into animals made it possible to explore the range of specificity which antibodies could show. It became clear that no limit could be found; even though antisera to one compound might react with another of related structure, it always appeared possible to absorb out the cross-reacting antibodies and obtain specific sera. Indeed, the increasing use of immunological techniques in all fields of biology and medicine to identify, estimate, and relate different substances in tissues and their extracts has depended on the ability to prepare functionally specific antisera, and few failures have been reported. It follows that antibody combining sites must be able to distinguish very slight differences in structure, and there are many examples of this from Landsteiner's work and subsequent work. It is perhaps worth mentioning one of relevance to the structure of immunoglobulins themselves. Inherited differences in antigenic specificity were detected in human IgG and were shown to be due to the presence of two allotypes, Inv 1,2 and Inv 3 (Ropartz *et al.*, 1961). Subsequent structural studies have shown that the only chemical difference on which this antigenic specificity depends is a change of a valine for a leucine residue in each of the two light chains of the IgG molecule (Baglioni *et al.*, 1966; Milstein, 1966), which has a molecular weight of 150,000. Unless the replacement causes a substantial configurational change, and this seems unlikely, it is a most remarkable example of the subtle structural differences which antibodies can detect. It has been argued, in support of successive theories of the mechanism of antibody formation, that the range of antibody specificity is much less than is generally believed, but there seems little basis to support this view.

Present knowledge of protein structure makes it certain that the specificity of an antibody depends on the arrangement of the amino acid residues in the combining sites, which in turn depends on the primary sequence of the sections of peptide chain involved. How sufficient variation of sequence could occur within a stable structure was difficult to envisage until the observation of Hilschmann and Craig (1965) that light chains have a variable and a constant section and the subsequent finding that this is true also of heavy chains. There is no doubt that this phenomenon provides the structural basis of antibody specificity, and rapidly accummulating data give increasing information about the nature of the site.

Ideas as to the size and general features of combining sites have been obtained indirectly by what has proved to be the easier approach of establishing the essential features of the antigenic determinants which complement the antibody site. More direct information is now being obtained by the use of reagents which can be localized in a combining site by their specific affinity for these sites and which subsequently react to form covalent bonds with the side chains of the amino acid residues in the site. Hydrolysis and characterization of labeled peptides identify the sequences in or near the site.

Knowledge of the antibody combining site will not be complete until the three-dimensional structure is established by high-resolution X-ray crystallography. A coherent picture can be obtained from the chemical studies. The interpretation of these is helped by our knowledge of the molecular structure of enzymes, obtained from crystallographic data, sometimes with competitive inhibitors held in the catalytic site (Porter, 1970).

II. SIZE AND GENERAL FEATURES OF THE ANTIBODY COMBINING SITE

The smallest fragment of antibody which retains full affinity for a hapten is the Fab fragment of 45,000 mol. wt. (Porter, 1960), and hence no direct evidence has been obtained of the size of the antibody combining site by attempts to split the molecule further.

In contrast, much progress has been made in investigating the size of the antigenic determinants of proteins, polysaccharides, and haptens. These studies have been reviewed by Karush (1962), Kabat (1966), and Crumpton (1966) (see also Schechter *et al.*, 1970). The common approach has been to identify fragments of the protein or polysaccharide in which affinity for the antibody can be demonstrated, usually by their ability to inhibit combination of antibody and whole antigen. Comparison of the inhibitory power of overlapping fragments of increasing size has shown that tetra- to heptapeptides and also saccharides and aromatic haptens of comparable size appear to have maximal inhibitory ability and hence presumably are equivalent to a complete antigenic site. Working with rabbit antidextran antibodies, Schlossmann and Kabat (1962) have shown clearly that in this as in other respects antibodies are a complex population of molecules with a range in the size of combining sites. The sites of IgM appear by this method to be smaller than those of IgG (Kaplan and Kabat, 1966); however, when antibodies to polyalanyl determinants were studied, no difference in the size of the combining site of IgM and IgG antibodies could be found (Haimovitch *et al.*, 1969).

Benjamini *et al.* (1969) have shown, however, that the whole of an antigenic determinant need not necessarily be concerned in the specificity of the interaction. In earlier work, it was found that rabbit anti-tobacco mosaic virus (TMV) could bind a decapeptide representing the sequence 103–112 of the TMV peptide chain. The minimum-length peptide which showed specific binding was five to seven residues from position 106, 107, or 108, to position 112, varying with the individual serum used. In all cases, the *C*-terminal sequence Leu-Asp-Ala-Thr-Arg was essential and no demonstrable binding of Ala-Thr-Arg with any antiserum was observed. However, specific affinity was regained if the octanoyl group was attached to this tripeptide. Affinity was estimated by comparison under standard conditions of the relative binding of the ^{14}C-labeled acetyl decapeptide and the ^{14}C-labeled octanoyl tripeptide. With seven different sera, the affinity was equal or higher with the latter compared to the former. The octanyl dipeptide (Thr-Arg) had no detectable affinity. It was essential that this last point be shown, as nonspecific binding of hydrophobic molecules by IgG has been reported by Parker and Osterland (1970). In this study, 8-anilinonaphthalene-1-sulfonate (ANS) was found to bind to rabbit IgG, to antibodies of various specificities, and to human and mouse myeloma proteins. The binding was predominately to the Fab fragments, with barely demonstrable affinities found for the Fc, and there was one binding site per Fab fragment. Affinity constants ranged from 10^3 liter per mole^{-1} to more than 10^4. The affinity varied from one myeloma protein to another and was significantly higher for antibody to the dinitrophenyl (DNP) group than for other immunoglobulins. This suggested that some type of cross-reaction was being measured and hence that the binding of the ANS was likely to be to the antibody combining site. However, competitive inhibition of binding between DNP-valine and ANS did not relate directly to the independent affinity constants, and it could not be concluded that the binding of the ANS molecules was to the antibody combining sites of the IgG. In particular, inhibition of binding of DNP-valine to anti-DNP IgG could be effected equally well by ANS and octanoic acid. As there is no structural relationship between the two inhibitors, both might be bound to a distinct hydrophobic site with a consequential conformational change of the antibody binding site. An apparent conformational change in serum albumin on the binding of 2 moles of fatty acid has been shown by Glazer and Sanger (1963), and Green (1963) has shown that fluorodinitrobenzene (FDND) can also be absorbed preferentially onto the same hydrophobic sites.

It can be concluded that an antigenic determinant is of the order of size of a hexapeptide or a hexasaccharide. A smaller determinant may be sufficient to establish the specificity of binding, but the affinity would be too small to demonstrate convincingly. The binding can be increased markedly by addition of a hydrophobic group such as a hydrocarbon chain. However, nonspecific binding of hydrophobic molecules can be demonstrated, and while this mimics specific binding in that there is one site per Fab fragment, the association constants rarely exceed 10^4. It is not clear whether these nonspecific binding sites are the same as the antibody binding sites. If they were, the increased affinity on attachment of hydrophobic groups to fragments carrying antigenic determinants would be due to a combination of specific and nonspecific forces and would imply that the antibody site contains hydrophobic regions. This would be in agreement with the observation that the affinity constants of antipolysaccharide antibodies for polysaccharides are generally lower than those of aromatic haptens for their antibodies.

It is therefore probable that antigenic determinants are of the same order of size as the substrates of an enzyme such as lysozyme. This enzyme recognizes a hexasaccharide, and hence the antibody combining site would be of similar size to the active-site cleft of lysozyme. In the 12 or 13 enzymes whose three-dimensional structure is known, the active centers are clefts or grooves lined with some 15-20 amino acid residues which may be in direct contact with the substrate.

It is reasonable to suppose, therefore, that the chemical determination of the antibody binding site will lead to the identification of 10-20 amino acid residues in the Fab region whose configuration and nature determine specificity.

These contact amino acids in the enzymes are not sequentially adjacent in the polypeptide chains, and it seems certain that this is true also of the amino acids in the antibody site. It is most probable, for instance, that the residues from both heavy and light chains are involved, and the effect of the presence of a hapten in stabilizing the structure of its antibody implies that different sections of the chains contribute. Cathou and colleagues (Cathou and Haber, 1967; Cathou and Werner, 1970; Cathou et al., 1968), using isolated rabbit anti-DNP antibodies of high affinity for the hapten ϵ-DNP-lysine (K_0 10^8), found that 4 M guanidine HCl did not dissociate any ϵ-DNP-lysine from the antibody and that 7.5 M guanidine HCl was necessary for the complete removal of the hapten. Moreover, as judged by circular dichroism or the quenching of fluorescence, the presence of the hapten had a marked stabilizing effect on the steric structure of the antibody. After 6 hr in 4 M guanidine HCl, the antibody in the absence of hapten appeared to have lost its native structure, but in the presence of the hapten only partial loss occurred. It was concluded that the interaction of the hapten and combining site was holding sections of the chain in their original position and that it was most likely that the effect was apparent because these sections were derived from linearly distant parts which on dissociation led to a substantial change of steric structure.

III. STRUCTURAL STUDIES

From the above considerations, it would be expected that the nature of the 10-20 residues in each Fab fragment would account for the combining specificity of the molecule and that these critical positions are likely to be scattered in the variable sections of both the heavy and light chains. As both these variable sections are in the Fab fragment, it seems certain that this is the mechanism by which antibody specificity is determined. Any contribution by the constant region seems unlikely because, while there are many alternatives here—κ or λ types of light chains and γ, a, μ, σ, or ϵ classes of heavy chains or their subclasses—no constant correlation of specificity with type, class, or subclass has been found. In certain circumstances, a given specificity may be associated with one type of constant region only—the most striking example perhaps being in the cold agglutinin disease, in which an antibody agglutinates red cells carrying the I antigen. Almost all cold agglutinins have only κ light chains, but exceptions with λ light chain have been observed; there is a striking but not absolute correlation between specificity and light-chain type (Harboe *et al.*, 1965).

Similar arguments apply to the allelic variants of both heavy and light chains. In individual heterozygous animals, a given specificity may be associated predominately with only one of the two allelic forms present, but in a population this is never a consistent feature, implying that the amino acid sequences determining the allelic specificity are not related to those determining combining specificities. As most of the sequences have been found in the constant sections of the heavy and light chains, such as the Inv and Gm of human immunoglobulins, this is not surprising. The only exceptions so far are the allelic specificities found in the heavy chains of rabbit immunoglobulin. These specificities are also unique in being shared among classes of heavy chain, and this observation has given strong support to the view that two genes code for each heavy chain. Here, allotype-related sequences are found scattered throughout the variable region at positions 10, 13, 15, 16, 17, 27, 28, 29, and 33 and also 80, 81, 82, 83, 84, and 85 (Wilkinson, 1969; Fruchter *et al.*, 1970; Fleischman, 1971; Mole *et al.*, 1971). This clearly implies that if all specificities can be found in antibodies of each heavy-chain allotype, then these residues play no part in determining combining specificity. Also, as there is no evidence that combining of antibody and antigen blocks allotypic specificity, it seems that these residues must be placed in such a position that their combination with antiallotypic sera does not obstruct the antibody combining site.

It is apparent, therefore, that there are constant areas in the variable sections which play no direct part in determining the specificity of the combining site. This is to be expected if not more than 20 positions of the 200 or more in the two variable sections of the chains of a Fab fragment are likely to be directly involved.

IV. SEQUENCE OF THE VARIABLE REGIONS OF HEAVY AND LIGHT CHAINS

The data available have been obtained predominately from human Bence-Jones proteins of the κ type together with some of λ type. Several sequences of mouse Bence-Jones proteins are also available, and the complete sequences of the variable regions of four human γ chains and one μ chain have been published. Much other more fragmentary data are available, particularly of N-terminal sequences and of sequences around the disulfide bonds. The complete sequence of both the heavy and light chains of one human myeloma IgG is known.

All this information has been summarized and analyzed at different stages of completeness by many authors, the most recent being a review by Milstein and Pink (1970) and a statistical analysis by Wu and Kabat (1970), who have used the sequences published for the light chains of human and mouse myeloma proteins. Perhaps the most informative analysis, given by Wu and Kabat (1970), is a plot of variability of residue against position for all light-chain sequences then available of both species. Variability in this case is expressed as

$$\frac{\text{Number of different amino acids at a given position}}{\text{Frequency of the most common amino acid at that position}}$$

In position 7, data on 63 chains were available, and four different residues (Pro, Thr, Ser, and Asp) have been reported. The most common, Ser, occurs in 41 positions, i.e., frequency 41/63 or 0.65 and variability 4/0.65 or 6.15.

This plot (Fig. 1) shows peaks at positions 28, 50, and 96, with much less

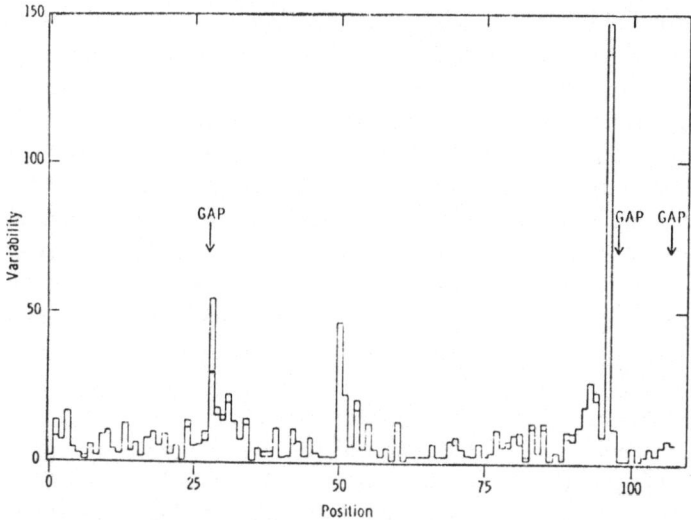

Figure 1. The variability at different amino acid residue positions for the variable region of the light chain (Wu and Kabat, 1970).

but relatively high variability in the adjacent amino acids. Gaps and insertions in the sequence also occur near 28 and 96, and a polysaccharide chain has been found attached to residues near position 28 in several Bence-Jones proteins. Polysaccharide has also been reported near position 60 in a human myeloma γ chain (Press and Hogg, 1970). Weigert *et al.* (1970) have reported the partial sequences of ten mouse λ chains and found six to be identical in the *N*-terminal 113 residues and four to differ in one or more of five positions: 25, 32, 50, 52, and 97 (i.e., in the three areas known previously to be the most variable). The exact numbering varies with the reference protein taken; the constant disulfide bridge in the variable region is given as positions 23-88 by Wu and Kabat (1970) and as 22-90 by Weigert *et al.* (1970).

All this information has been used principally as a basis for theories of the genetic origin of this remarkable variability, but for the present purposes all that needs to be emphasized is that the short sections of hypervariability, within the variable region, are likely to be the parts of the chain involved directly in the combining site.

The much more limited data on the variable sections of the heavy chains show comparable features, with exceptional variability in two areas *C*-terminal to the two half-cystines of the intrachain disulfide bond. The third hypervariable section, between positions 50 and 60, may become apparent as more data accumulate. The information to date shows that the greatest variability is in positions 95-96 in the light chain (Fig. 1) and 100-110 in the heavy chain—the variability of the other two hypervariable sections being significantly lower than these. Although sufficient sequences of the heavy-chain variable sections are not yet available to be certain, this most variable section seems to include some five or six positions in the heavy chain but only one or two in the light chain.

Equally striking is the absolute constancy of certain parts of the variable sections—most notably the intrachain disulfide bonds. Though numbering may vary, depending on the insertions and deletions elsewhere in the chain, the intrachain disulfide bond is always present in the same position with almost constant adjacent amino acids. If the allotype-related amino acids in the rabbit γ chain do indeed control allotypic specificity, then they also must be constant for a given allotype. It is noteworthy that the hypervariable regions appear to be so close in both cases to the constant structural features.

The sequences of the variable region of the heavy chain of the rabbit IgG of allotypic sequences Aa1, Aa2, and Aa3 (Fig. 2) emphasize this point. Because of the half-cystine residue in position 92, a radioactive label S-carb-oxymethyl-^{14}C could be introduced, and this could be used to assess the recoveries of the peptides from tryptic digestion terminating in arginine at 94. This led to the conclusion that all or nearly all molecules in both the Aa1 and Aa3 IgG from pooled serum had the same sequence from positions 81 to 94 and that that from 86 to 94 was identical in both allotypes (Mole *et al.*, 1971). Few recognizable peptides from positions 95 to about 115 could be found, presum-ably due to the very high variability in this section and hence very low yield of any one peptide. Support for this was obtained by preparing radioactive peptide maps of peptides from a chymotryptic digest all commencing Cys-Ala-Arg and continuing into the hypervariable region. Some 30 peptides of different mobil-ities were recognized. In rabbit IgG, this most variable section follows a sequence apparently common to all molecules, and this in turn follows a hexapeptide sequence unique to each allotype.

Kabat (1967) has drawn attention to some 29 positions which appear to be almost constant in both human κ and λ chains and emphasized that seven are occupied by glycine, which he suggests could be of particular significance because the absence of a side chain in this residue facilitates flexibility and accommodation of changes in other residues. Two invariant glycine residues at positions 99 and 101 are well placed in, or close to, the most variable section. However, reference to known structures shows that adjacent residues fill spaces due to the absence of glycine side chains, and there appears to be no particular flexibility about the position of this residue. Glycine is one of the most stable residues among proteins such as hemoglobin and cytochrome from different species (Dayhoff, 1969). Koshland and colleagues have carried out a careful series of analyses of rabbit antibodies to haptens, both cross-reacting and non-cross-reacting (Koshland *et al.*, 1966, 1969, 1970). After allowance was made for the effects due to class, subclass, and allotype, reproducible differences

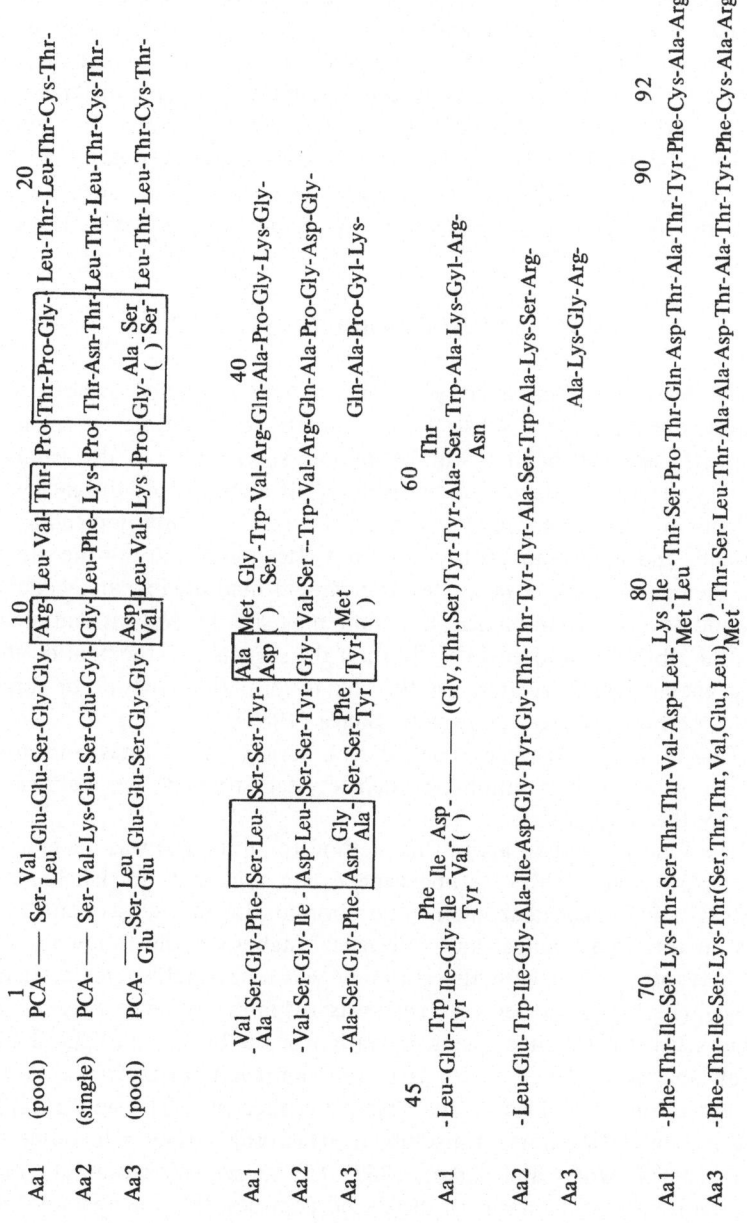

Figure 2. Sequences of the variable region of the heavy chains of rabbit IgG of allotypes Aa1, Aa2, and Aa3 (Mole *et al.*, 1971). Allotype-related sequence changes are enclosed in a box.

were found among antibodies of different specificities. When, for example, the amino acid analyses of anti-lac and antiarsonic antibodies were compared, there were clear-cut differences between them: six aspartic acid, seven serine, two or three alanine, two valine, and six tyrosine residues. The same differences were found in antibodies of the IgG and of the IgM fractions. In contrast, if antibodies against cross-reacting haptens and arsonic and phosphoric acids were compared, the analyses differed in only one tyrosine residue. Sequence data to support and position these differences found by analysis have not yet been reported.

V. AFFINITY-LABELING STUDIES

The principle of the affinity-labeling method as applied to both enzymes and antibodies is that a substrate, a competitive inhibitor, or an antigenic determinant such as hapten will be bound specifically to the catalytic or combining site of an enzyme or an antibody. If a group able to react with the side chains of the amino acids is substituted into such low molecular weight compounds and if the substitution does not alter the affinity of the site for the ligand, then subsequent covalent reaction should bind the low molecular weight compound to residues in or near the combining site. Hydrolysis and identification of the labeled peptides will establish, if the sequence of the peptide chains is already known, which sections of the chains form the catalytic or combining site. The method has been reviewed by Singer (1967).

This technique has been applied to a variety of enzymes, and in several cases the subsequent solution of their crystal structure has confirmed the validity of the approach.

The first attempt to apply this method to antibodies was made by Wofsy *et al.* (1962) using rabbit anti-benzenearsonic acid and the labeling reagent *p*-(arsonic acid)benzenediazonium fluoroborate. In the subsequent series of papers on this topic, Singer and colleagues established the following: (1) The initial rate of reaction with the antibody was considerably greater than with inert IgG. (2) This enhanced reactivity was lost in the presence of excess hapten (arsonic acid). (3) On subsequent separation of the heavy and light chains, the ratio of label was about 2:1 for H/L. (4) The reaction occurred entirely with tyrosine residues, as judged by the absorption spectrum. The only exception to this was found using horse antiserum to β-lactoside, when a histidine residue became labeled (Wofsy and Parker, 1967). (5) Subsequent proteolytic digestion led to the isolation of several small labeled peptides, the principal one from the heavy chain being Thr–Tyr and from the light chain Val–Tyr, with the label in both cases being on the tyrosine residue.

Further extension of the method has been difficult because the sequence of the variable sections of the heavy and light chains of rabbit IgG was not known and the heterogeneity of the labeled peptides handicapped the isolation of peptides large enough to allow positioning by comparison with the known sequences from the peptide chains of human and mouse myeloma proteins. However, Franek (1971) has recently reported the isolation of two peptides from a digest of the light chain of pig antibody labeled with m-nitrobenzenediazonium fluoroborate. Antibodies had been isolated and labeled according to the techniques of Singer, the light chains isolated, and the κ and λ chains separated (Franek and Zorina, 1967). Two labeled sequences were identified from the λ chain and positioned by comparison with the partial sequence of the variable region of pig λ chain reported previously (Franek and Novotny, 1969). In both sequences, the label was bound to a tyrosine residue, one identified as occupying position 33 and the other position 93. The difficulties of working with complex sequences were avoided in part in this work by starting with large amounts of reacted antibody; thus 50 g antibody was used to provide 3 g of λ chain for enzymic hydrolysis.

These difficulties have been largely avoided by Goetz and Metzger (1970a,b), using the same reagent, m-nitrobenzenediazonium fluoroborate and a mouse myeloma IgA protein, MOPC 315, which had been shown to bind DNP-lysine with high affinity (Eisen et al., 1967, 1968). The protein is, of course, homogeneous, and equilibrium dialysis shows a uniform binding affinity over a wide concentration range. The light chain is of λ type, though unique in having Cys-Leu at the C-terminus, and sequence data on this and other λ chains from IgA myeloma proteins of mice of the same Balb/c strain are accummulating. When the labeling reagent was allowed to react with the MOPC 315 protein, it was found to be bound almost entirely to the light chain; 1.4 mole of label were reported per mole protein of mol. wt. 150,000, but there is some uncertainty about the molecular weight of this IgA myeloma protein, which may be higher than the 150,000 assumed. A tryptic peptide prepared from the light chain in good yield was sequenced and by its very close homology with the sequence of another Balb/c myeloma λ chain was identified as coming from positions 24-56, with the label substituted on a tyrosine residue in position 34. This is in a position comparable to that found by Franek (1971) working with the λ chain of pig anti-DNP antibodies. The homology between the sequences of the two labeled peptides is apparent (Fig. 3).

Mouse: -Val-Thr-Thr-Ser-Asp-Tyr(Ala,Ser)Trp-Ile-

Pig: -Val-Thr-Thr-Ser-Asn-Tyr-Pro-Gly-Trp-Phe-

Figure 3. Sequences from the λ chains of a mouse myeloma protein MOPC 315 and pig anti-DNP antibody reacting with ω-nitrobenzenediazonium fluoroborate. In both cases, the label has reacted with the tyrosine residue (Goetz and Metzger, 1970a,b; Franek, 1971).

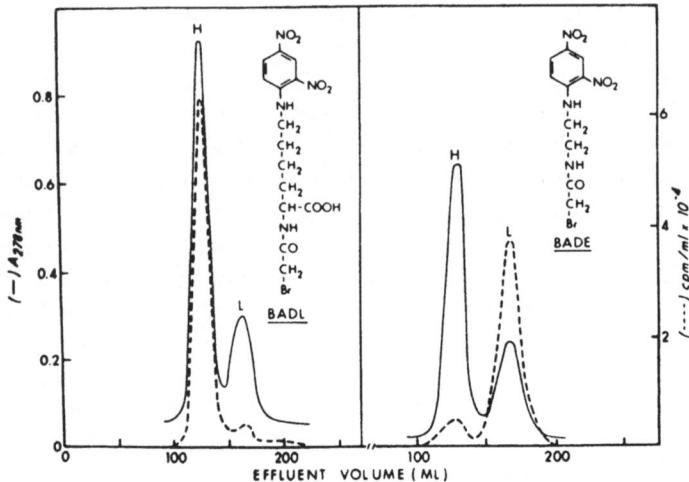

Figure 4. Affinity labeling of mouse IgA myeloma protein MOPC 315. The two reagents BADL and BADE react with MOPC 315. By subsequent reduction and separation of the peptide chains by chromatography in propionic acid, it was found that BADL was bound to the heavy chain while BADE was bound to the light chain (Haimovitch *et al.,* 1970).

A surprising feature of these experiments was the observation that while the affinity of the substituted protein DNP-caproate-^{3}H compared with that of the unsubstituted protein had fallen 50-fold there still appeared to be 1.4 binding sites per 150,000 mol. wt. Thus the reaction with the labeling reagent is specific, and has caused some change in the combining site, but does not occupy it after covalent reaction. The reagent may partially hinder access to the site or may have caused some indirect alteration.

The reaction of this same MOPC 315 myeloma protein with two other DNP labeling reagents has been reported (Haimovitch *et al.,* 1970). These were BADE and BADL, prepared with ^{14}C-labeled bromoacetic acid (Weinstein *et al.,* 1969; Givol *et al.,* 1970), and differed in that BADL is three carbons longer and carries a carboxyl group relative to BADE (see Fig. 4). Both these reagents reacted in a comparable way to *m*-nitrobenzenediazonium fluoroborate to give 1.4 mole bound per protein mol. wt. of 150,000. However, when the peptide chains were separated, the BADL label was bound to a lysine residue on the heavy chain and the BADE to a tyrosine residue on the light chain. No sequence data to position the labels more closely are available at present, nor is there information on the binding affinities of the labeled proteins. The difference in reaction of the two reagents is presumably due to the difference in length and charge and the availability of reactive amino acid residues in the neighborhood of the site. This work shows clearly the possibility of mapping the peptide sequences adjacent to the combining site and gives much essential information on the steric arrangement of the relevant parts of the peptide chain.

Some uncertainty was felt about the specificity of reaction of substances such as aryl diazonium salts and bromacetyl groups, since reaction with lysine and tyrosine residues not in the site was also possible. This appeared to occur in the reaction between rabbit antihapten antibodies and substituted hapten, since, even though the labeling rate was much greater with antibody than with inert IgG, it was necessary to allow only 0.5 mole of label to react, because nonspecific labeling became significant under more effective conditions. Franek (1971) reduced the nonspecific reactions by working in unfavorable conditions for reaction (pH 5, 4°C). The clearness of the results with the MOPC 315 myeloma protein, where 1.4 or more of sites are labeled with little evidence of nonspecific labeling, suggests that the difficulty with rabbit antibodies arose largely from the wide range of rates of reaction. The uniformly high affinity sites of MOPC 315 myeloma protein react sufficiently rapidly under the conditions used so that nonspecific substitution is negligible. It is apparent, however, that in working with proteins of lower affinity or mixtures of natural antibodies there would be advantage in using labeling reagents whose reactive group could be activated after binding in the combining site. Two such have been developed, a diazoketone (Converse and Richards, 1969) and an aromatic azide (Fleet et al., 1969).

Both are activated by light, the first giving rise to a carbene and subsequently a ketene (Fig. 5). The latter is able to react only with lysine, tyrosine,

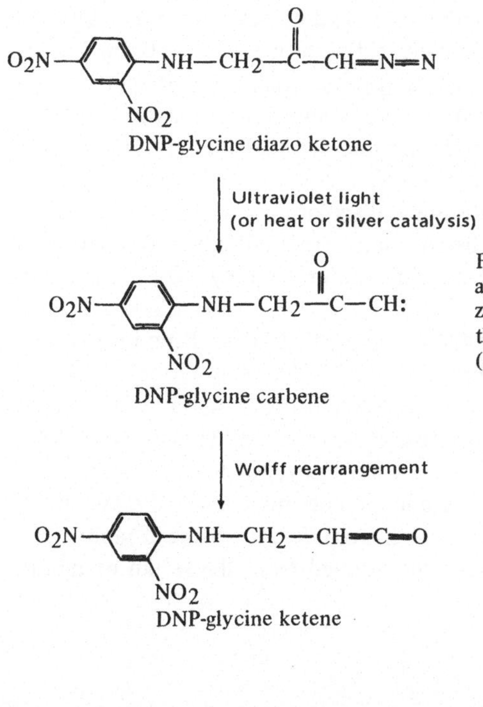

DNP-glycine diazo ketone

Ultraviolet light
(or heat or silver catalysis)

DNP-glycine carbene

Wolff rearrangement

DNP-glycine ketene

Figure 5. The activation by light of the affinity-labeling reagent DNP-glycinediazoketone. It seems likely that reaction with the protein would occur through the ketene (Converse and Richards, 1969).

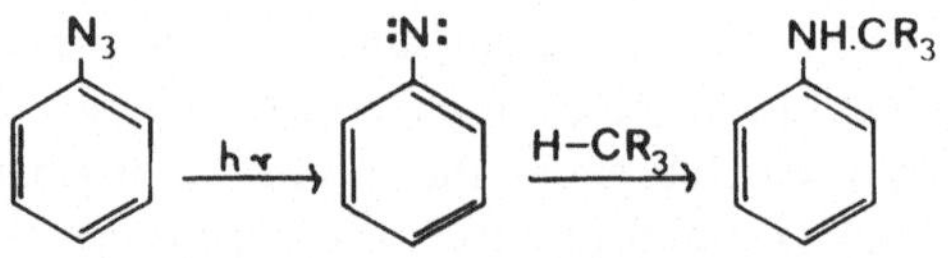

Figure 6. The activation by light of an aromatic azide to give a nitrene
able to insert into a C–H bond (Fleet *et al.*, 1969).

and histidine, as do the diazo and bromacetyl derivatives, but the carbene could insert into C–H bonds and hence react with any amino acid. It is not known which reaction predominates, as no labeled derivations have been characterized. The second, the aromatic azide, is converted to a nitrene (Fig. 6) able, as the carbene, to insert into any C–H, N–H, or S–H group; however, as no rearrangement is likely, the aromatic azide has the advantage that it should be able to react with any amino acid in the combining site. The limited range of reactivity of the other three reagents may be important, as a failure of a hapten carrying a diazo group to react with its antibody has been reported (Koyama *et al.*, 1968), and this may arise from the absence of any of the three reactive amino acid residues near the combining site. The failure of *m*-nitrobenzenediazonium fluoroborate to block fully the combining site of MOPC 315 protein after reaction at Tyr 34 on the light chain may be because this residue is so placed that reaction is not possible with the hapten still in position in the site.

The nitroazide reagent (NAP-lysine) has the additional advantage that it can be substituted on a protein in the dark and injected into a rabbit to stimulate antibodies against the potentially active group itself. Hence the reactive group becomes an antigenic determinant rather than being attached to the determinant. It would be expected that in these circumstances reaction with the contact amino acids of the binding site rather than with adjacent residues would occur.

In fact, all three reagents (diazonium, diazoketone, and aromatic azide) used with rabbit antibodies label the heavy chains preferentially with a ration of 2:1 to 4:1 for H/L, but in each case a maximum of only 50-60% covalent reaction could be achieved.

The failure to achieve nearly 100% covalent binding has been investigated further using the nitrene reagent (Press *et al.*, 1971), as isolation of labeled peptides (below) confirmed that this reagent can form N–C bonds and hence any amino acid side chain should be reactive. It was reported (Fleet *et al.*, 1969) that 1.1 mole of hapten was found per mole antibody, but in subsequent work the figures have lain between 0.6-0.9 mole hapten per mole antibody. Two kinds of antibody preparation have been used, each from pooled serum of rabbits of a single allotype (Aa3). One preparation was isolated from the serum of rabbits immunized with NAP bovine globulin and the other from rabbits immunized with DNP bovine globulin. In the latter case, some 70% of the antibody able to precipitate with DNP human serum albumin would precipitate with NAP human serum albumin, and this is the fraction of antibodies which was used. Such antibody preparations bound noncovalently 1.8-2 mole of hapten per mole antibody, with affinity constants between 5×10^5 to 7×10^6 liter per mole^{-1} at $4°C$, and if antibody and excess hapten were passed down a column of Sephadex G-25 in the dark, 2 mole hapten per mole antibody remained bound, If, however, light was shone on such a solution, only 0.8 mole hapten and 0.6 mole hapten per mole anti-DNP antibody and anti-NAP antibody, respectively, were bound covalently. When the reacted preparations were again mixed in the dark with NAP-lysine, sufficient was bound noncovalently to bring the total back to 1.8-2 mole, proving that the nonreacted sites were free. However, if light was now shined, no additional covalent reaction occurred.

If the reacted antibody was fractionated on an immunoabsorbent column, 25-40% antibody was not absorbed and contained 1.5-1.8 mole covalently bound hapten, while the absorbed antibody contained 0.2-0.4 mole. The latter would bind NAP-lysine up to 2 mole per mole antibody, but no reaction occurred on to exposure to light. Thus there appear to be two fractions of antibody of the same specificity each able to bind hapten noncovalently, but only in one could a covalent reaction follow activation of the reactive group by light. The affinity constants of the two fractions showed no significant difference.

The myeloma protein MOPC 315 with a high affinity for DNP-lysine showed no affinity for NAP-lysine, comparable to the 30% of the rabbit anti-DNP antibody which also showed no affinity for NAP-lysine. It would seem probable that if a myeloma protein with affinity for NAP-lysine were found, there would be a 2:1 probability that on activation it would not react covalently with the hapten. The orientation of the hapten in the binding site may decide whether (1) the nitrene inserts into an amino acid side chain to form a stable covalent bond, (2) forms some kind of very labile bond which is split on subsequent handling, or (3) reacts with solvent and not an amino acid. The orientation of the hapten is likely to be dependent on the precise structure of the binding site and hence to show all variations in any antibody preparation but only one in a myeloma protein.

The position may well be different if the reactive group of the affinity-labeling compound is not a determinant, as in the substituted diazonium salts and bromoacetyl derivatives. In these cases, the reaction may well not occur in the combining site, as no reactive amino acid side chain may be present. Reaction which is allowed to continue over hours could occur in the vicinity of the site, which would be functioning as a concentrating agent rather than a binding site. If this were true, the precise orientation of the reagent in the site would be unimportant, and complete reaction might be expected with all antibodies or myeloma protein binding the reagent. The difficulty in this case lies in the nonspecific reactions which can occur simultaneously.

Isolation and characterization of the labeled peptides from the light chain of pig antibodies reacted with *m*-nitrobenzenediazonium salt have been discussed above. With the light chain of rabbit antibodies to the DNP group, Thorpe and Singer (1969) reported the isolation of a dipeptide, Val-Tyr, which they suggested might be from a sequence immediately prior to the half-cystine at position 88, but Franek, working with pig antibodies, identified conclusively a nonapeptide in which a tyrosine residue several positions after the same half-cystine residue was labeled.

In the heavy chain of rabbit antibodies to the DNP group, Thorpe and Singer (1969) isolated a dipeptide, Thr-Tyr, with the tyrosine labeled. Press *et al.* (1971) isolated two labeled peptides, Cys-Ala-Arg and Phe-Cys-Ala-Arg, from the NAP antibodies in which the alanine and the half-cystine residues, respectively, were labeled. The sequence toward the end of the variable region of the heavy chain is (O'Donnell *et al.*, 1970)

$$\overset{*}{\text{Asp-Thr-Ala-Thr-}}\overset{90}{\text{Tyr-Phe-}}\overset{*}{\text{Cys-}}\overset{*}{\text{Ala-}}\overset{94}{\text{Arg}}$$

All these labeled peptides appear to be derived from this section, which is equivalent to the position in the light chain in which Franek (1971) found that the minor labeled peptide had originated.

In the rabbit heavy chain, the sequence of this section appears to be exceptionally constant in that positions 86-94 are identical in both Aa1 and Aa3 allotypes. Positions 80-85 are different in Aa1 and Aa3 but constant in the heavy chain of one allotype. This section of the chain cannot therefore determine specificity, but it is immediately adjacent to the most variable section (position 95 to about position 115). It is likely that the specificity is controlled by these very variable residues but that the orientation of the labeling reagent is such that reaction occurs with the neighboring stable residues. If reaction is occurring also with the variable residues, the isolation and characterization of the labeled peptides will be difficult, as any one peptide may well be in very low yield.

VI. CONCLUSION

All three lines of approach (physical studies, sequence data, and affinity labeling) are giving consistent results in that they suggest that the combining sites of antibodies are formed from a small number of amino acid residues which are scattered through both the heavy and light chains. This is suggested for the following reasons:

1.   The presence of hapten by its noncovalent binding stabilizes the steric structure against disruption by denaturing agents such as guanidine.

2.   There are several hypervariable positions C-terminal to the half-cystines of the intrachain disulfide bond in the variable region of both heavy and light chains, with possibly another hypervariable region about midway between. This is true of both chains, but at present there appear to be rather more hypervariable positions in the heavy chain.

3.   In the affinity-labeling studies where sufficient sequence data are available, one position (Tyr 34) C-terminal to Cys 22 in the light chain of both pig antibody and a myeloma protein has been labeled. Several residues adjacent to Cys 94 in the heavy chain of rabbit IgG including the Cys 94 itself also have been labeled. With the same myeloma protein, either heavy or light chains may be labeled, depending on the reagent used, and with rabbit antibodies the labels were found on both chains but preferentially on the heavy. It is therefore probable that the amino acid residues lining the combining site are in positions 30 ± 2 and 96 ± 2 on the light chain and 31-32 and 95-105 on the heavy chain. Position 50 in the light chain and possibly an equivalent in the heavy chain may also be present. There is evidence that hydrophobic residues may be adjacent to the site.

4.   A noteworthy feature is the presence of very stable sequences near to these hypervariable residues. They are near Cys 23 and 88 in the light chain and 22 and 94 in the heavy chain. If the allotype-related sequences 27-29 and 85-90 in the rabbit γ chain are indeed responsible for the allotypic specificity, then they must be absolutely constant in antibody of given allotype, and, as mentioned earlier, no allotypic restriction of specificity has been shown. It might be suggested that these stable features are the rigid framework onto which the variable combining site is secured. There are other stable residues contributing to this structure such as the allotype-related residues in rabbit heavy chain and the subgroup-determining residues found in the human light chains.

5.   The practicability of such a model may be considered against the known structures of several enzymes and their combining sites. All such sites are clefts or grooves lined with 10-20 residues. In the model of lysozyme, it is apparent that some of these residues could be changed with no effect on the general structure of the protein. Aspartic acid residue 101 is an obvious example of such a residue which projects into the cleft and which could therefore be changed with little effect on the structure of the molecule. It is believed to form a bond with the substrate, and any change would have a pronounced effect on the specificity of binding. Another contact residue is aspartic acid 52, but in this case its position is much more restricted by its closeness to other sections of

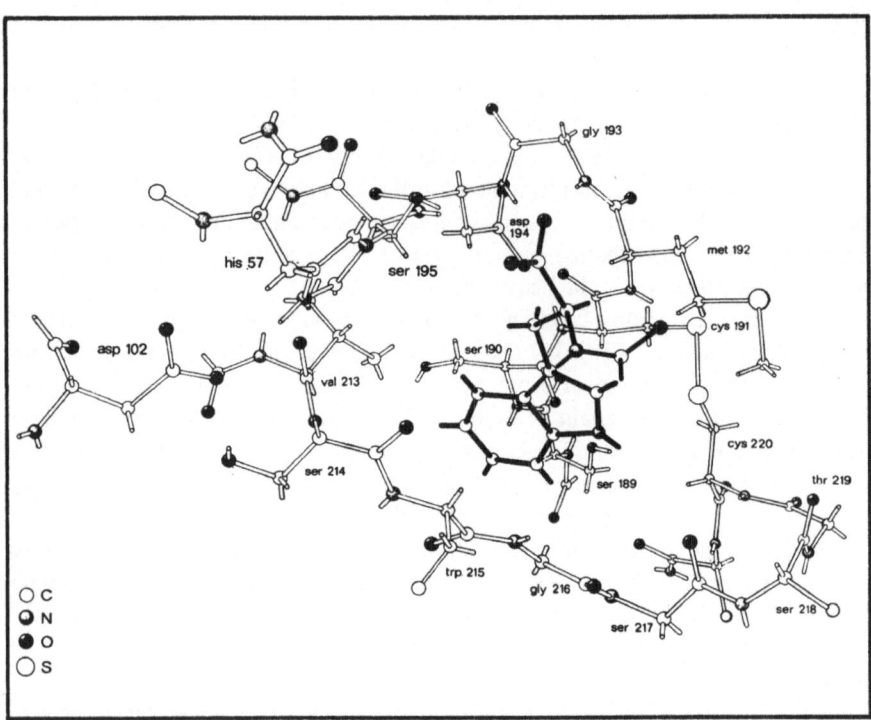

Figure 7. Stereo drawing of the active site of a-chymotrypsin with formyl-1-tryptophan in the position deduced from the electron density map. The numbers indicate the positions of the residues in the linear sequence (Steitz *et al.*, 1969).

chain, and any change would probably necessitate other compensatory changes if the overall structure of the molecule were to be retained.

Just how easily the specificity of binding of small molecules by protein can be changed can be seen from the comparison made by Hartley (1970) of the catalytic sites of chymotrypsin (Birktoft *et al.*, 1970) and elastase (Shotton and Watson, 1970) and the postulated site of trypsin. These sites (Fig. 7) differ only in that serine 189 in chymotrypsin is changed to an aspartic acid in trypsin. In elastase, the corresponding gap (tosyl hole) into which an aromatic or basic residue fits is blocked by substitution of a valine for glycine 216. These small changes in an otherwise similar structure alter the specificity of binding from that of a peptide containing an aromatic or leucine residue in chymotrypsin to that of a peptide containing a basic residue in trypsin or to much less specific binding of any peptide in elastase.

From consideration of these enzyme structures, there is no difficulty in envisaging an antibody combining site as a cleft where the changing of a 15-20 residue can alter the specificity of binding in a great variety of ways. Coupled with some flexibility in the shape of the cleft, it is likely that a very wide range of specific affinities could be accommodated.

Heterogeneity of antibodies is, of course, a most characteristic feature, and, however careful the isolation procedure, any preparation of antibodies against a hapten such as DNP-lysine will be a mixture of antibodies with a range of affinity constants and a range of cross-reactions with related compounds. This implies slight differences in the structure of the site, different orientation of binding of hapten, and the possibilities of slight as well as substantial changes in the residues and the positions of the residues lining the site. Again, this is not difficult to picture against the background of the known enzyme structures. There are many changes of residue in the variable region outside these highly variable positions postulated to be actually in the site. These other changes are likely to be necessary to accommodate alteration in the site of a residue such as Asp 52 of lysozyme. There are obviously possibilities of the same residue being in a slightly shifted position depending on alteration elsewhere. Examination of models leaves little doubt that the possibilities of variation in such a system are likely to be as great as the range of antibody specificity demands. Only determination of the steric structure of an immunoglobulin molecule will complete our knowledge of an antibody combining site, but continuation of the chemical studies discussed above should clarify many features of this remarkably efficient method of building sites with a very wide range of specificity and affinity within a constant framework.

REFERENCES

Baglioni, C., Aleseio Zonta, L., Cioli, D., and Carbonava, A. (1966). *Science* 152:1519.
Benjamini, E., Shimizu, M., Young, J. D., and Lenng, C. Y. (1969). *Biochemistry* 8:2242.
Birktoft, J. T., Blow, D. M., Henderson, R., and Steitz, A. (1970). *Phil. Trans. Roy. Soc.* 257:67.
Cathou, R. E., and Haber, E. (1967). *Biochemistry* 6:513.
Cathou, R. E., and Werner, T. C. (1970). *Biochemistry* 8:3149.
Cathou, R. E., Kulezyeki, A., and Haber, E. (1968). *Biochemistry* 7:3958.
Converse, C. A., and Richards, F. F. (1969). *Biochemistry* 8:4431.
Crumpton, M. J. (1966). In Cinader, B. (ed.), *Antibodies to Biologically Active Molecules*, Pergamon Press, Oxford.
Dayhoff, M. O. (1969). *Atlas of Protein Sequence and Structure*, Vol. 14, National Biomedical Research Foundation.
Eisen, H. N., Little, J. R., Osterland, C. K., and Simms, E. S. (1967). *Cold Spring Harbor Symp. Quant. Biol.* 32:75.
Eisen, H. N., Simms, E. S., and Potter, M. (1968). *Biochemistry* 7:4126.
Fleet, G. W. J., Knowles, J. R., and Porter, R. R. (1969). *Nature* 224:511.
Fleischman, J. F. (1971). *Biochemistry* 10:2753.
Franek, F. (1971). *Europ. J. Biochem.* 19:176.
Franek, F., and Novotny, J. (1969). *Europ. J. Biochem.* 11:165.
Franek, F., and Zorina, O. M. (1967). *Collection Czech. Chem. Commun.* 32:3229.
Fruchter, R., Jackson, S. A., Mole, L. E., and Porter, R. R. (1970). *Biochem. J.* 116:249.

Glazer, A. N., and Sanger, F. (1963). *J. Mol. Biol.* 7:452.

Givol, D., Weinstein, Y., Gorecki, M., and Witche, K. M. (1970). *Biochem. Biophys. Res. Commun.* 38:825.

Goetz, E. J., and Metzger, M. (1970a). *Biochemistry* 9:1267.

Goetz, E. J., and Metzger, M. (1970b). *Biochemistry* 9:3862.

Green, N. M. (1963). *Biophys. Biochim. Acta* 74:542.

Haimovitch, J., Schechter, I., and Sela, M. (1969). *Europ. J. Biochem.* 7:567.

Haimovitch, J., Givol, D., and Eisen, H. N. (1970). *Proc. Natl. Acad. Sci.* 67:1656.

Harboe, M., Van Forth, R., Schubothe, H., Kind, K., and Evans, R. S. (1965). *Scand. J. Haematol.* 2:259.

Hartley, B. S. (1970). *Phil. Trans. Roy. Soc.* 257:77.

Hilschmann, N., and Craig, L. C. (1965). *Proc. Natl. Acad. Sci.* 53:1403.

Kabat, E. A. (1966). *J. Immunol.* 97:1.

Kabat, E. A. (1967). *Proc. Natl. Acad. Sci.* 58:229.

Kaplan, M. E., and Kabat, E. A. (1966). *J. Exptl. Med.* 123:1061.

Karush, F. (1962). *Adv. Immunol.* 2:1.

Koshland, M. E., Engleberger, F. M., and Shapanka, R. (1966). *Biochemistry* 5:641.

Koshland, M. E., Davis, J. J., and Fujita, N. J. (1969). *Proc. Natl. Acad. Sci.* 63:1274.

Koshland, M. E., Ochoa, P., and Fujita, N. J. (1970). *Biochemistry* 9:1880.

Koyama, J., Grossberg, A. L., and Pressman, D. (1968). *Biochemistry* 7:1935.

Milstein, C. (1966). *Nature* 209:370.

Milstein, C., and Pink, J. R. (1970). *Progr. Biophys. Mol. Biol.* 21:209.

Mole, L. E., Jackson, S. A., Porter, R. R., and Wilkinson, J. M. (1971). *Biochem. J.* (in press).

O'Donnell, I. J., Frangione, B., and Porter, R. R. (1970). *Biochem. J.* 116:261.

Parker, C. W., and Osterland, C. K. (1970). *Biochemistry* 9:1074.

Porter, R. R. (1960). *Brookhaven Symp. Biol.,* No. 13, p. 203.

Porter, R. R. (1970). Harvey Lecture.

Press, E. M., and Hogg, N. M. (1970). *Biochem. J.* 117:641.

Press, E. M., Fleet, G. W. J., and Fisher, C. E. (1971). In Amos, B. D. (ed.), *Progress in Immunology,* Academic Press, New York.

Ropartz, G., Lenoir, J., and Rivat, L. (1961). *Nature* 189:586.

Schechter, B., Schechter, I., and Sela, M. (1970). *J. Biol. Chem.* 245:1438.

Schlossmann, S. F., and Kabat, E. A. (1962). *J. Exptl. Med.* 116:535.

Shotton, D. M., and Watson, H. C. (1970). *Phil. Trans. Roy. Soc.* 257:111.

Singer, S. J. (1967). *Adv. Protein Chem.* 22:1.

Steitz, T. A., Henderson, R., and Blow, D. M. (1969). *J. Mol. Biol.* 46:337.

Thorpe, N. O., and Singer, S. J. (1969). *Biochemistry* 8:4523.

Weigert, M. G., Cesari, I. M., Yomkovich, S. J., and Cohn, M. (1970). *Nature* 228:1045.

Weinstein, Y., Wilchek, M., and Givol, D. (1969). *Biochem. Biophys. Res. Commun.* 35:694.

Wilkinson, J. M. (1969). *Biochem. J.* 112:173.

Wofsy, L., and Parker, D. C. (1967). *Cold Spring Harbor Symp. Quant. Biol.* 32:111.

Wofsy, L., Metzger, H., and Singer, S. J. (1962). *Biochemistry* 1:1031.

Wu, T. T., And Kabat, E. A. (1970). *J. Exptl. Med.* 132:211.

γD Immunoglobulin

Hans L. Spiegelberg*

Department of Experimental Pathology
Scripps Clinic and Research Foundation
La Jolla, California

I. INTRODUCTION

In the past 15 years, much has been learned about the structure and function of different classes of antibodies from studies of monoclonal immunoglobulins formed in patients with multiple myeloma and macroglobulinemia. This is particularly true for γD, the fourth class of human immunoglobulins. γD was discovered as the result of the study of an unusual human myeloma protein (Rowe and Fahey, 1965a), and all subsequent investigations of γD have depended in one or another phase on the availability of γD myeloma proteins. Two characteristics of γD often make it necessary to study γD myeloma proteins rather than normal γD. First, the concentration of γD in the serum is very low, an average of 30 μg/ml, and, second, γD is a labile immunoglobulin which has a great tendency to aggregate and to fragment into subunits during the isolation procedure. γD can therefore usually not be isolated in sufficient quantities from normal blood. Probably for this reason, progress in the evaluation of the role of γD in the immune mechanism of the body has been slow. Although the striking similarity of the structure of γD myeloma proteins to that of the other classes of immunoglobulins has always suggested that γD must represent a class of human antibodies, several years passed following its discovery before antibody activity in that class was reported; to date, all antibody activity has been demonstrated by indirect methods. Therefore, definite proof of antibody activity in isolated normal human γD still remains to be shown. Since γD does not share any of the biological activities of the other immunoglobulin classes, such as fixation of

* Faculty Research Associate of the American Cancer Society.

complement and induction of anaphylactic reactions in the skin, role evaluation remains difficult. It is the purpose of this chapter to review the present knowledge of human γD immunoglobulin and to stimulate new investigations which, it is hoped, will lead to discovery of the biological role of γD antibodies.

II. ISOLATION OF γD

γD is more difficult to isolate than all the other immunoglobulins. First, γD is susceptible to proteolysis and is fragmented by plasmin and probably other proteolytic enzymes in the serum into Fab and Fc fragments during the isolation procedure. Second, the physicochemical properties of γD are very similar to those of γA, and, since there is about 100 times more γA than γD in the serum, relatively large quantities of γD can only rarely be obtained from normal serum. Therefore, γD has usually been isolated from patients with γD multiple myeloma, whose sera contain large amounts of γD and only small quantities of γA.

The same guidelines can be followed for the isolation of both normal and myeloma γD except for a last step, which is necessary to isolate normal γD. Serum or plasma is collected, and aminocaproic acid (EACA) is immediately added to a final concentration of 0.02 M (262 mg/100ml). EACA is added in order to prevent the formation of plasmin from plasminogen (Alkjaersig *et al.*, 1959), apparently the proteolytic enzyme mainly responsible for the "spontaneous" fragmentation of γD that occurs during the storage and the isolation of γD. Once plasmin is formed, EACA no longer prevents fragmentation of γD. The serum or plasma is stored at -20°C or, if the facilities are available, at -70°C. in order to prevent plasmin formation during the isolation procedure, 0.01 M EACA is added to a final concentration of 0.01 M (1.31 g/liter) to *all* buffers used.

The isolation of γD myeloma proteins usually involves two steps: DEAE-cellulose chromatography or block electrophoresis followed by Sephadex G-200 exclusion chromatography. When DEAE–cellulose chromatography is used as the first step, the serum is dialyzed against 0.015 M phosphate buffer (*p*H 8.0) and applied to a DEAE-cellulose column equilibrated with the same buffer. Most of the γG is eluted from the column with this buffer, and the γD is subsequently eluted with 0.035 M and 0.05 M phosphate buffer (*p*H 8.0). γD myeloma proteins having an electrophoretically slow mobility are eluted from the column with the 0.035 M buffer and those with a more anodal mobility with 0.05 M buffer. Instead of stepwise elution, a gradient elution of the γD can also be used. Additional EACA (approximately 0.01 M) is added to the serum fractions immediately following elution of the DEAE-cellulose column, since γD appears to be most susceptible to fragmentation at this point.

The serum can also be fractionated in the first step by Pevikon (Müller-Eberhard, 1960) or starch (Kunkel, 1954) block electrophoresis. The fractions containing the γD are concentrated by pressure dialysis and applied in the cold to a Sephadex G-200 or Biogel P-200 column equilibrated with pH 7.0 phosphate-buffered 0.15 M NaCl containing 0.01 M EACA. The γD is eluted from the column between the peak containing γM and the peak containing γG. Preparations of γD myeloma protein obtained in this manner are usually quite pure and contain only traces of γA and other β proteins. If, however, γD is isolated from normal serum, relatively large quantities of γA and other serum β proteins contaminate the γD fraction, and further purification is required. This step consists of passing the fraction rich in γD through an immunoadsorbent column which is made of insolubilized anti-γA antibodies and, if desired, also of insolubilized anti-whole serum antibodies that have been absorbed with purified γD myeloma proteins. Of the many procedures for the formation of immuno-adsorbents, three which have been used successfully in our laboratory are the insolubilization of rabbit or goat antibodies with ethylchloroformate (Avrameas and Ternynck, 1967) and the coupling of the antibodies to either Sepharose 4B (Cuatrecasas et al., 1968) or bromacetylcellulose (Robbins et al., 1967).

III. STRUCTURE OF γD

A. Physical Properties

1. Sedimentation Rate

Sedimentation constants have been reported for a total of seven individual γD myeloma proteins by a number of laboratories (Rowe et al., 1969; Hanson et al., 1966; Spiegelberg et al., 1970b). The $S^{\circ}_{20,w}$ values (S) reported vary from 6.14 to 7.04, with an average of 6.55. As shown by Rowe et al. (1969), the S rate of γD appears to be less dependent on concentration than does that of γG. The reason for variations of S rates of myeloma proteins is not known; however, different techniques used by various laboratories do not seem to account solely for the variations. Significant differences in the S rates of three myeloma proteins were found in the same laboratory by Hanson et al. (1966). As these authors also found differences in the S rates of individual γG myeloma proteins, this may be characteristic of all myeloma proteins. Possibly, some differences in S rates are the result of variable carbohydrate contents of individual proteins and of degrees of contamination of the γD preparations. Polymer formations characteristically seen in γA myeloma proteins and in γM macroglobulins have, however, not been reported for γD myeloma proteins.

2. Molecular Weight

The molecular weight of γD myeloma proteins has been determined by approach to sedimentation equilibrium (Yphantis, 1964). The values reported by three laboratories are 172,000 (Spiegelberg *et al.*, 1970*b*), 184,000 (Rowe *et al.*, 1969), and 200,000 (Saha *et al.*, 1970). The molecular weight of the γD heavy (δ) chain has also been studied, by both exclusion chromatography and polyacrylamide gel electrophoresis in sodium dodecyl sulfate (SDS). By use of exclusion chromatography on Sephadex G-200 columns equilibrated with 0.05 M formic acid containing 8 M urea, a molecular weight of 60,000 was obtained for the δ chain (Spiegelberg *et al.*, 1970*b*). Examination of the δ chain by SDS-polyacrylamide gel electrophoresis yielded a molecular weight of 70,000 (Perry and Milstein, 1970). Causes for the discrepancy are not clear. Differences of molecular weights determined by sedimentation equilibrium could result from assumption of incorrect values of the partial specific volume ($\bar{v}$), since the $\bar{v}$ has as yet not been determined for γD. The $\bar{v}$ has been calculated at 0.717 for one myeloma protein from the amino acid and carbohydrate composition (Spiegelberg *et al.*, 1970*b*).

It is interesting to note that if the lower molecular weights are assumed to be correct and if the 15% carbohydrate of the δ chain is taken into consideration, the molecular size of the peptide portion of the δ chain is similar to that of the γ chain—50,000. In contrast, an additional pseudounit of about 120 amino

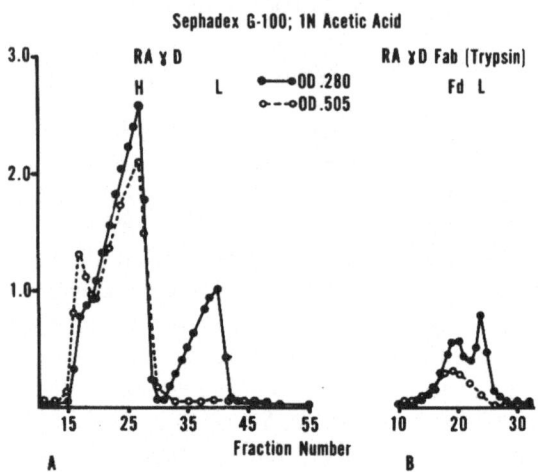

Figure 1. A: Separation of heavy and light chains of a reduced and alkylated γD myeloma protein (RAγD). The carbohydrate, as measured by the orcinol method, is present in the heavy chain. B: Separation of Fd fragment and light chain from reduced and alkylated tryptic Fab fragment (RAγD Fab). The carbohydrate is present on the Fd fragment, which is eluted from the column before the light chain.

acids has to be postulated if the higher molecular weights obtained for the δ chain should represent the correct value (Perry and Milstein, 1970).

3. Extinction Coefficient

The extinction coefficient, $E^{1\%}_{280nm, 1cm}$, was determined for four isolated γD myeloma proteins dissolved in pH 7.0 phosphate-buffered 0.15 M NaCl in our laboratory (Spiegelberg, unpublished data). The percent of protein nitrogen of γD was calculated from the amino acid and carbohydrate composition to be 14.5%, and the protein nitrogen content of the γD preparations was determined by an automated micro-Kjeldahl method. Based on these values, extinction coefficients of 14.5, 16, and, twice, 17.5 were obtained for four myeloma proteins. Addition of 0.01 M EACA to the buffered saline did not change the extinction coefficient. If the extinction coefficient of a γD myeloma protein is not known, it appears best to assume the average $E^{1\%}_{280nm, 1cm}$ of 17.0.

4. Heat and Acid Lability

γD resembles γE in its heat and acid lability and differs from γG, γA, and γM in these properties. Heiner et al. (1968) reported that the precipitability of γD by a specific antiserum was lost to 50% following heating for 1 hr at 56°C and 90% following incubation for 4 hr. Complete loss was observed following incubation for 4 hr at 61°C. γD was also denatured following exposure of acid pH, e.g., to 0.1 M glycine buffer (pH 3.0) and to 2 M KCl (pH 4.0). This is important to know if one attempts to isolate γD from an immunoadsorbent. γD cannot be eluted from an immunoadsorbent column by acid treatment, but only by chaotropic agents such as 2 M KCl or 2 M KI at pH 7.4 (Heiner et al., 1968).

B. Chemical Properties

1. Polypeptide Chains

Like all other immunoglobulins, γD can be separated into heavy and light polypeptide chains following reduction and alkylation of the interchain disulfide bonds (Fig. 1). γD was reduced in tris buffer (pH 8.2) in the presence of 0.1 M β-mercaptoethanol or 0.02 M dithiothreitol (DTT) for 1-2 hr at room temperature and subsequently alkylated with 0.11 or 0.05 M iodoacetamide, respectively. The reduced and alkylated γD was dialyzed against 1 N acetic acid and applied to a Sephadex G-100 column equilibrated with 1 N acetic acid. The first peak (OD_{280}) eluted from the column contained the heavy (δ) chain and the second peak the light chain. As indicated from the recovery of protein

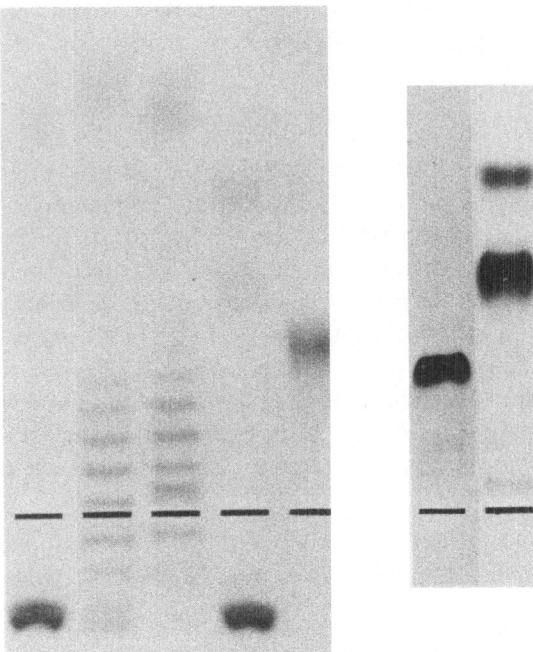

Figure 2. Starch gel electrophoresis employing a glycine buffer (pH 8.8), anode on top (1-5) and an 8 M urea formic acid buffer (pH 3.5), anode on bottom (6, 7). (5) and (6) γD (May). (1) γD digested for 1 hr with papain in the presence of reducing agent; (2) γD digested with trypsin for 2 min; (3) "spontaneously" split γD; (4) γD digested with papain for 1 hr in the absence of reducing agent; (7) reduced and alkylated γD.

nitrogen in the heavy- and light-chain fractions and from the mass ratio calculated from the respective molecular weights, γD is composed of two heavy and two light chains.

Separation of γD into heavy and light polypeptide chains can also be demonstrated by urea starch gel electrophoresis. As shown in Fig. 2, the native γD appears as a single band, whereas the reduced and alkylated γD separates into two faster-moving bands. The four polypeptide chains are linked by three interchain disulfide bonds, two inter-heavy-light and one inter-heavy-heavy chain disulfide bonds. This conclusion is based on data of the uptake of ^{14}C-labeled iodoacetamide and of the recovery of carboxymethylcysteine (CMCys) in heavy and light chains which had been reduced and alkylated under conditions where no reduction of intrachain disulfide bonds occurred (Spiegelberg et al., 1970b). This is further substantiated by the finding of only two labeled peptides obtained by peptide mapping (Spiegelberg *et al.*, 1970*b*) and paper electro-phoresis (Perry and Milstein, 1970), each containing 1 mol of CMCys per mole.

2. Amino Acid Composition and Sequence Analysis

In general, the amino acid content of the δ chain is unremarkable as compared to that of γ chains (Spiegelberg *et al.*, 1970*b*). The presence of 10 mole of CMCys per mole of fully reduced and alkylated δ chain suggests that the δ chain contains four intrachain disulfide bonds in addition to the two interchain disulfide bonds. The δ chain of one γD myeloma protein was shown to contain 8 mol of methionine per mole of δ chain, three methionine residues being localized to the Fc fragment (Spiegelberg, unpublished data). Saha *et al.* (1970) reported a low proline content for two γD myeloma proteins, whereas the proline content of three γD myeloma proteins studied in our laboratory was similar to that of γG: 122, 126, and 130 mole per mole.

Relatively little is known about the amino acid sequence of the δ chain. No free amino (*N*) terminal amino acid has been found in three δ chains, and no carboxy (*C*) terminal amino acid residue could be liberated by treatment of the δ chains with carboxypeptidases A and B, even in 2 M urea (Spiegelberg *et al.*, 1970*b*). Fragmentation of the δ chain with cyanogen bromide gave rise neither to the formation of the *C*-terminal octadecapeptide characteristic of γ chains nor to the octapeptide characteristic of *a* and *μ* chains (Spiegelberg *et al.*, 1970*b*). Amino acid sequence data have recently been reported by Perry and Milstein (1970) for the peptides containing the inter-heavy-light and inter-heavy-heavy chain disulfide bonds (Fig. 3). The amino acid sequence of the peptide containing the inter-heavy-light chain disulfide bond shows a relatively good homology to the analogous peptides in human γG$_2$, γG$_3$, γG$_4$, mouse γ$_{2b}$, and rabbit γG, which suggests that the light chain of γD is linked to the δ chain in a position similar to that of these proteins about 100 residues *N*-terminal of the inter-Fd–Fc region. The peptide containing the inter-heavy-heavy chain disulfide bond does not show a striking homology to any of the known amino acid sequences of the other immunoglobulin classes.

3. Carbohydrate Contents

γD myeloma proteins contain about 12% carbohydrate, which is usually all bound to the δ chain (Spiegelberg *et al.*, 1970*b*; Perry and Milstein, 1970). The carbohydrate consists of about 40 mole of hexoses (D-galactose and D-man-nose in a ratio of 1:1), 20 mole of *N*-acetylglucosamine, 10 mole of *N*-acetyl-galactosamine, and 6 mole of *N*-acetylneuraminic acid per mole of δ chain, and very little, if any, fucose. Higher sugar values have been reported by Saha *et al* (1970) for two γD myeloma proteins, one of them having carbohydrate linked λ   . . . -Pro-Thr-Glu-Cys-Ser-

δ . . . -Pro-Ile-Ser-Gly-Cys-Arg- . . . . -Thr-Pro-Glu-Cys-Pro-Ser-His-Thr-Gln-Pro-Leu-Gly- Val- . . . .

δ . . . -Pro-Ile-Ser-Gly-Cys-Arg- . . . . -Thr-Pro-Glu-Cys-Pro-Ser-His-Thr-Gln-Pro-Leu-Gly- Val- . . . .

λ   . . . -Pro-Thr-Glu-Cys-Ser-

γ2,γ3,γ4 -Pro-Leu-Ala-Pro-Cys-Ser-Arg-

Figure 3. Amino acid sequence of peptides of the δ chain containing the interchain disulfide bonds
(Perry and Milstein, 1970).

to the light chain. It is possible, therefore, that a small percentage of the γD myeloma proteins contain carbohydrate on the light chain, similar to γG myeloma proteins (Spiegelberg *et al.*, 1970*a*).

The carbohydrate of the δ chain is distributed among at least three glycopeptides (Spiegelberg *et al.*, 1970*b*). One contains galactosamine as its characteristic amino sugar and is located in the inter-Fd–Fc (hinge) region. The other two contain glucosamine as the sole amino sugar, and it is localized in the Fc fragment.

4. Proteolytic Fragments

In the study of the first recognized γD myeloma protein, Rowe and Fahey (1965*a*) showed that γD, like γG, could be fragmented by papain into an electrophoretically fast migrating fragment (Fc) which contained the antigenic determinants characteristic of the γD class, and into a slowly migrating fragment (Fab) which contained the antigenic determinats of the light chains. These data were extended to include results from the digestion of γD with the enzymes trypsin, plasmin, and pepsin and a modification of the papain treatment omitting reducing agents (Spiegelberg *et al.*, 1970*b*). The fragments obtained from γD following "spontaneous" degradation were also characterized. In these experiments, it was noted that prolonged digestion with any of these enzymes resulted in degradation of the Fc fragment and, in some cases, even the Fab fragment. However, relatively well-defined fragments could be obtained in high yields following brief digestion periods, and it could be demonstrated that different enzymes split γD at readily distinguishable sites (Figs. 2 and 4).

Digestion of γD with papain for 1 hr at 37°C in the absence of a reducing agent resulted in the formation of Fab fragments which did not contain carbohydrate, as all the carbohydrate of γD was recovered in the Fc fragment (Figs. 2 and 4). Fab fragments lacking carbohydrate were also obtained following digestion of γD with papain in the presence of reducing agents; however, the Fc fragment isolated following this procedure contained only about two thirds of the carbohydrate, and a relatively large quantity of small peptides were liberated.

In contrast to papain digestion, digestion of γD with 1% trypsin for 2 min at *p*H 8.0 yielded Fab fragments which contained about one third of the carbohydrate of γD and, in particular, all the galactosamine (Figs. 2 and 4). The tryptic Fc fragment contained about two thirds of the carbohydrate, including all the glucosamine of γD. Little, if any, peptide material was released following digestion with trypsin for this short period of time. Besides the difference in the carbohydrate content, the fragments obtained following papain and trypsin digestion differed also in their electrophoretic mobility, as shown by starch gel electrophoresis (Fig. 2). The papain Fab fragment appeared as a single band

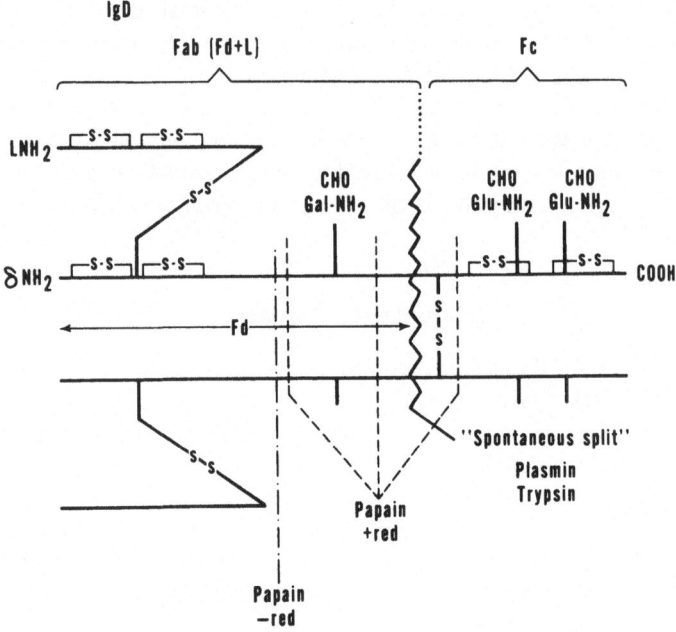

Figure 4. Schematic model of the γD molecule based on present information of the structure of γD. Gal-N, glycopeptide containing galactosamine as characteristic amino sugar; Glu-N, glycopeptide containing glucosamine as the sole amino sugar.

moving toward the cathode, whereas the tryptic Fab fragment showed multiple bands. The multiple banding was shown to be the result of heterogeneity in the sialic acid content of individual molecules, since treatment of the tryptic Fab fragment with neuraminidase reduced the number of bands to one having an electrophoretic mobility similar to that of the papain Fab fragment. The tryptic Fc fragment was characterized by a more anodal electrophoretic mobility than the Fc fragment obtained following digestion with papain in the absence of reducing agent. The Fc fragment obtained following digestion with papain in the presence of reducing agent had a mobility similar to that of the tryptic Fc fragment. No antigenic difference could be demonstrated between Fab or Fc fragments obtained by either papain or trypsin digestion.

Fragmentation of γD with plasmin and the "spontaneously" occurring proteolysis of γD yielded fragments which are very similar, if not identical, to those obtained following brief digestion with trypsin.

Digestion of γD with pepsin for 18 hr resulted in the formation of primarily dialyzable peptides. Shorter periods yielded a heterogeneous mixture of fragments which could not be studied because of paucity of material.

The tryptic Fab fragment can be separated following reduction and alkylation into Fd fragment and L chain, probably because of the carbohydrate moiety on the Fd fragment (Fig. 1).

5. Schematic Model of the γD Molecule

A schematic model based on the present knowledge of the structure of γD is shown in Fig. 4. γD is composed of two heavy and two light polypeptide chains, which are linked by three interchain disulfide bonds. The inter-heavy-light chain disulfide bond is localized in the Fab fragment, and the inter-heavy-heavy chain disulfide bond in the Fc fragment. The exact positions of the interchain disulfide bonds are unknown; however, homology of the sequence to the peptide containing the inter-heavy-light bond to the analogous peptide of other immunoglobulin classes suggests its location to be approximately 100 residues N-terminal from the inter-Fd–Fc (hinge) region. In addition to the interchain disulfide bonds, the δ chain contains four intrachain disulfide bonds, two each in the Fd and Fc fragments.

γD contains about 12% carbohydrate, which is distributed among at least three glycopeptides. One glycopeptide containing galactosamine as the characteristic amino sugar is localized in the inter-Fd–Fc region, and two glycopeptides containing glucosamine as the sole amino sugar are located in the Fc fragment. Papain, in the absence of a reducing agent, fragments γD on the N-terminal side of the glycopeptide containing galactosamine. Trypsin, plasmin, and the "spontaneously" occurring fragmentation split γD on the C-terminal side of this glycopeptide. Digestion of γD in the presence of reducing agent results in fragmentation at multiple sites, even after only brief periods of digestion.

The most striking difference between γD and the other classes of immunoglobulins appears to be the enzymatic cleavage into Fab and Fc fragments. In contrast to γG, which is digested by papain and trypsin at a very similar site (Rutishauser et al., 1968), the γD is digested by these enzymes at readily distinguishable sites, as shown by the absence or presence of carbohydrate on the respective Fab fragments. Furthermore, γD is very susceptible to proteolysis and fragments "spontaneously" unless it is stored in the presence of EACA. The reason for this great susceptibility to proteolysis is presently unknown.

Although the structure of γD taken as a whole is quite distinct from that of other immunoglobulins, certain individual structural features show striking similarities to one or another class of immunoglobulins. The high carbohydrate content of γD is similar to that found in human γM and γE. The glycopeptide containing galactosamine localized in the inter-Fd–Fc region appears in a position homologous to one found thus far only in human γA_1 (Abel and Grey, 1967; Ko et al., 1967) and rabbit γG (Smyth and Utsumi, 1967). The presence of a single inter-heavy-heavy chain disulfide bond is unique among human immunoglobulins but has been shown in rabbit γG (Palmer and Nisonoff, 1964). No free

N-terminal amino acid was demonstrable in the few δ chains thus far examined, which suggests that the variable region of the δ chains resembles the two major variable heavy-chain subgroups (Cunningham *et al.*, 1969).

IV. BIOLOGICAL PROPERTIES OF γD

A. γD Concentration in Body Fluids

Of the many immunoglobulins, γD is a relatively minor component, representing only about 0.25%. The concentration of γD in the serum of normal individuals varies greatly, and the normal population can be divided into three groups, according to γD concentration (Rowe and Fahey, 1965*b*). The major group (about 70%) has a concentration of 20-50 μg of γD per milliliter of serum. The second group (about 15%) is characterized by a very low γD concentration of less than 3 μg per milliliter, and the third group shows a very high concentration of 100-400 μg or more per milliliter. γD is first detected at about 6 months of age; older children vary in the γD concentration, as do adults (Rowe *et al.*, 1968*b*). The reason for the large differences in γD serum concentration is not known. A genetic factor does not appear to be responsible, since differences have also been observed in identical twins (Rowe *et al.*, 1968*a*); however, such studies are incomplete. Possibly, γD antibodies are formed to only a few antigens, and high concentrations of γD are observed in individuals sensitized with those antigens. The great concentration difference is not unique for γD; it has also been observed for another minor immunoglobulin component, the subclass γG_4 (Kunkel *et al.*, 1970).

γD concentrations have been studied in a wide variety of diseases (Rowe *et al.*, 1968*b*). As in the normal population, the γD concentration varies greatly in all diseases studied, making statistically significant data difficult to obtain. In chronic infections, γD levels are increased (Rowe *et al.*, 1968*c*), as are those of all other immunoglobulins, but to date no specific increase of γD has been associated with a particular disease. Patients with allergies and autoimmune diseases do not show an abnormal γD concentration, and γD is usually absent in hypogammaglobulinemic individuals (Rowe *et al.*, 1968*b*). γD has not been detected in normal colostrum, saliva, lacrimal fluid, jejunal fluid, bile, urine, or cerebrospinal fluid (Rowe *et al.*, 1968*b*). By use of the fluorescent antibody technique, occasional plasma cells staining for γD can be shown in the spleen, lymph nodes, and adenoid tissue. However, no significant increase of γD-positive cells has been noted in the various organs (Pernis *et al.*, 1966).

B. Antibody Activity

Although γD was first recognized in 1965 as the fourth class of human immunoglobulines, antibody activity associated within this class has not been reported for some time. Moreover, all data of activity described to date have been obtained by indirect antibody- or antigen-binding tests. Direct proof of activity in a purified γD preparation, therefore, still remains to be shown. The major problems which one faces in demonstrating γD antibody activity are the presence of only minute quantities of γD antibody in immune sera and the lack of knowledge of a unique biological function of γD which could be used to design sensitive biological tests. γD antibodies to several antigens were, however, demonstrated by extremely sensitive indirect methods. By use of the fluorescent antibody technique, γD antinuclear antibodies were shown in about 40-50% of the sera of patients with systemic lupus erythematosus and in 20% of sera of patients with rheumatoid diseases (Ritchie, 1968; Watson *et al.,* 1969; Kantor *et al.,* 1970). Utilizing the same method, Kantor *et al.* (1970) described antithyroid antibodies in one of 26 patients with Hashimoto's disease. Heiner and Rose (1970) demonstrated γD antibodies to bovine serum albumin and bovine gamma globulin in three of 20 sera from patients with cow's milk sensitivity and to diphtheria toxoid in the serum of an immunized person by radio-Ouchterlony analysis. Utilizing a hemagglutination-enhancing technique, Gleich *et al.* (1960) demonstrated γD antibodies to penicillin. γD anti-insulin antibodies were shown by a red blood cell linked antigen-antiglobulin reaction by Devey *et al.* (1970). These authors attempted to isolate, at least partially, the γD from the serum, but were unsuccessful in demonstrating activity in the isolated γD preparations. The titer of all γD antibodies reported was very low, and this might be the reason for the unsuccessful attempts to isolate γD antibodies. Furthermore, if normal γD, like myeloma γD, is susceptible to proteolysis, all future attempts to isolate normal γD antibodies should include EACA added to the serum and isolation buffers in order to minimize the spontaneous degradation of γD. Attempts to demonstrate γD antibodies in sera obtained from rhesus monkeys immunized with keyhole limpet hemocyanin and bovine serum albumin were unsuccessful in our laboratory.

C. Metabolic Properties

Rogentine *et al.* (1966) studied the metabolism of γD in normal individuals and in patients with an abnormal immunoglobulin metabolism. Low γD serum concentration was demonstrated to result primarily from a rate of synthesis (0.4 mg/kg/day) which is 100 times lower than that of γG. In addition, γD has a fast catabolic rate. The serum half-life is about 3 days, and the fractional turnover rate is 37%. The synthetic rate differed greatly from person to person, whereas the catabolic rate was similar. Seventy-three percent of the γD equilibrated into the intravascular compartment as compared to 44% of γG. γD appears to be metabolized independently of the other immunoglobulins.

D. Other Biological Activities

To date, neither a biological activity unique for γD nor an activity shared with other classes of immunoglobulins has been demonstrated. γD does not fix complement (Henney et al., 1969), neither the first component (Spiegelberg, unpublished data) nor late components as do human γA immunoglobulins (Spiegelberg et al., 1972) and guinea pig γ_1 antibodies (Sanford and Osler, 1971). γD does not sensitize guinea pig skin for reverse passive cutaneous anaphylaxis (Henney et al., 1969; Ovary, 1969). No release of lysosomal enzymes from neutrophils was observed following addition of aggregated γD myeloma proteins, whereas γG and γA myeloma proteins caused significant release (Henson et al., 1972). That γD is not transferred through the placenta is demonstrated by its absence in cord serum (Rowe et al., 1968b).

E. γD Multiple Myeloma

γD multiple myeloma is relatively uncommon. The incidence reported varies from 0.6 to 3% of all myeloma patients studied (Fahey et al., 1968; Fishkin et al., 1970; Zawadzki and Edwards, 1967). Patients with γD myeloma have several clinical and immunochemical characteristics that can be distinguished from those of other forms of multiple myeloma. The concentration of the γD paraprotein is usually lower than that of γG, γA, and γE myeloma proteins, probably because of the rapid catabolism of γD. The light-chain type of γD myeloma proteins is λ in 80-90% of the cases, and almost all patients have heavy Bence-Jones proteinuria. Furthermore, it appears that extraosseous tumors are more frequent in γD myeloma and that a younger age group is affected (Hobbs and Corbett, 1969). The median survival following diagnosis is about two years, similar to that in other forms of myeloma.

V. PROSPECTS

The question "What is the function of γD immunoglobulins?" has not been answered to date, 6 years following their discovery by Rowe and Fahey (1965a). Almost certainly, γD represents a class of human antibodies. In structure, γD resembles closely other classes of immunoglobulins, and γD antibody activity, although demonstrated only by indirect methods, has been reported to several antigens. The biological significance of γD, however, is completely unknown. Several difficulties were encountered in studying the function of γD. First, the quantity of γD antibodies found in the serum was very low, and attempts to isolate these antibodies were unsuccessful. Second, γD does not share a biological activity with any other class of immunoglobulins. Third, no specific increase of γD has been found related to any disease, and, fourth, an immunoglobulin class analogous to human γD has not been detected in animals, limiting all experimental studies.

What research can and should be performed in order to learn more about the role of γD? Obviously, possession of human sera containing a relatively large amount of γD antibodies which could be isolated, purified, and studied would be helpful. To find such sera, γD serum concentration should be determined regularly in large numbers of patients. Histories of patients having unusually high γD concentrations should then be reviewed carefully for clues to antigens which could have elicited a γD response, and the serum should be analyzed for the presence of γD antibodies using a panel of antigens. Also, such patients and volunteers should be immunized with various vaccines to determine their ability to form γD antibodies.

Another approach to the study of the biological function of γD is suggested by the striking similarity of γD to γE antibodies. Both γE and γD are present in the serum in very low concentrations, and both are heat labile. γE, like γD, is rapidly catabolized and does not fix complement. It has been shown that γE is cytophilic for basophils and mast cells and, upon reaction with antigen, initiates the release of vasoactive substances from these cells (Ishizaka, 1970). It appears from these studies that cell-bound γE is biologically far more important than the serum γE. It is possible that γD is also cytophilic to a specific cell and functions in conjunction with this cell. Careful analysis of the cytophilic properties of γD to white blood cells known to participate in an immune response would test this hypothesis.

Continuing studies of the primary structure of the δ chain will lead to accurate information as to its size and amino acid sequence. The degree of homology of the δ chain to other heavy chains can then be determined, offering insight into the structural and functional relationship between γD and other immunoglobulin classes. Since many of these studies depend on the availability of rare γD myeloma proteins, physicians having patients thus afflicted and making the plasma available could help investigators involved in the study of the structure and function of γD immunoglobulins.

ACKNOWLEDGMENT

This is Publication No. 525 from the Department of Experimental Pathology, Scripps Clinic and Research Foundation, La Jolla, California. The work was supported by American Heart Association Grant 70-710, American Cancer Society California Division Special Grant 531, and Atomic Energy Commission Contract AT (04-3)-410.

REFERENCES

Abel, C. A., and Grey, H. M. (1967). *Science* 156:1609.
Alkjaersig, N., Fletscher, A. P., and Sheiry, S. (1959). *J. Biol. Chem.* 234:832.
Avrameas, S., and Ternynck, T. (1967). *J. Biol. Chem.* 242:1651.
Cuatrecasas, P., Wilchek, M., and Anfinsen, C. B. (1968). *Proc. Natl. Acad. Sci.* 61:636.
Cunningham, B. A., Pflumm, M. N., Rutishauser, U., and Edelman, G. M. (1969). *Proc. Natl. Acad. Sci.* 64:997.
Devey, M., Sanderson, C. J., Carter, D., and Coombs, R. R. A. (1970). *Lancet* 2:1280.
Fahey, J. L., Carbone, P. P., Rowe, D. S., and Bachmann, R. (1970). *Am. J. Med.* 45:373.
Fishkin, B. G., Glassy, F. J., Hattersley, P. G., Hirose, F. M., and Spiegelberg, H. L. (1970). *Am. J. Clin. Pathol.* 53:209.
Gleich, G. J., Bieger, R. C., and Stankievic, R. (1969). *Science* 165:606.
Hanson, U. B., Laurell, C. A., and Bachmann, R. (1966). *Acta Med. Scand.Suppl.* 445:89.
Heiner, D. C., and Rose, B. (1970). *J. Immunol.* 104:691.
Heiner, D. C., Saha, A., and Rose, B. (1968). *Fed. Proc.* 27:489.
Henney, C. S., Welscher, H. D., Terry, W. D., and Rowe, D. S. (1969). *Immunochemistry* 6:445.
Henson, P., Bell, H. J., and Spiegelberg, H. L. (1972). *Clin. Exp. Immunol.* (in preparation).
Hobbs, J. R., and Corbett, A. A. (1969). *Brit. Med. J.* 1:412.
Ishizaka, K. (1970). *Ann. Rev. Med.* 21:187.
Kantor, G. L., van Herle, A. J., and Barnett, E. V. (1970). *Clin. Exptl. Immunol.* 69:951.
Ko, A., Clamp, L. R., Dawson, G., and Cebra, J. (1967). *Biochem. J.* 105:35P.
Kunkel, H. G. (1954). *Meth. Biochem. Anal.* 1:140.
Kunkel, H. G., Joslin, F. G., Penn, G. M., and Natvig, J. B. (1970). *J. Exptl. Med.* 132:508.
Müller-Eberhard, H. J. (1960). *Scand. J. Clin. Lab. Invest.* 12:33.
Ovary, Z. (1969). *J. Immunol.* 102:790.
Palmer, J. L., and Nisonoff, A. (1964). *Biochemistry* 3:863.
Pernis, B., Chiappino, G., and Rowe, D. S. (1966). *Nature* 211:424.
Perry, M. B., and Milstein, C. (1970). *Nature* 228:934.
Ritchie, R. F. (1968). *Rheumatism* 11:506.
Robbins, J. B., Haimovich, J., and Sela, M. (1967). *Immunochemistry* 4:11.
Rogentine, G. N., Rowe, D. S., Bradley, J., Waldmann, T. A., and Fahey, J. L. (1966). *J. Clin. Invest.* 45:1467.
Rowe, D. S., and Fahey, J. L. (1965a). *J. Exptl. Med.* 121:171.
Rowe, D. S., and Fahey, J. L. (1965b). *J. Exptl. Med.* 121:185.
Rowe, D. S., Boyle, J. A., and Buchanan, W. W. (1968a). *Clin. Exptl. Immunol.* 3:233.
Rowe, D. S., Crabbe, P. A., and Turner, M. W. (1968b). *Clin. Exptl. Immunol.* 3:477.
Rowe, D. S., McGregor, I. A., Smith, S. J., Hall, P., and Williams, K. (1968c). *Clin. Exptl. Immunol.* 3:63.
Rowe, D. S., Dolder, F., and Welscher, H. D. (1969). *Immunochemistry* 6:437.
Rutishauser, U., Cunningham, B. A., Bennet, C., Konigsberg, W. H., and Edelman, G. M. (1968). *Proc. Natl. Acad. Sci.* 61:1414.
Saha, A., Chowdhury, P., Sambury, S., Behelak, Y., Heiner, D. C., and Rose, B. (1970). *J. Immunol.* 105:238.
Sanford, A., and Osler, A. (1971). Complement Workshop, Baltimore.
Smyth, D. G., and Utsumi, S. (1967). *Nature* 216:332.
Spiegelberg, H. L., Abel, C. A., Fishkin, B. G., and Grey, H. M. (1970a). *Biochemistry* 9:4217.
Spiegelberg, H. L., Prahl, J. W., and Grey, H. M. (1970b). *Biochemistry* 9:2115.
Spiegelberg, H. L., Gotze, O., and Muller-Eberhard, H. J. (1972). *Fed. Proc.* (in press).
Watson, I., Heiner, D., Rose, B., and Bootello, A. (1969). *Clin. Res.* 17:362.
Yphantis, D. A. (1964). *Biochemistry* 3:297.
Zawadzki, Z. A., and Edwards, G. A. (1967). *J. Clin. Pathol.* 48:418.

Index

Actinomycin D, 120, 122, 129, 130
Adrenocorticotropic hormone
  as an antigen, 3, 36
  delayed hypersensitivity to, 36, 37
  determinant location, 32
  structure–function relationship, 32
Affinity-labeling, 154–161
  p-(arsonic acid) benzenediazonium fluoro-
    borate, 154
  m-nitrobenzenediazonium fluoroborate,
    155, 156, 158
  of antibody combining site, 158–161
  principle of method, 154
Alloantibodies, 66
Alloantigens
  chemistry, 54
  HL-A determinants, 68
  specificity ratios, 66
  solubilization, 54–55
Allotypes
  chemical differences, 145–146
  Inv groups, 145
Aminocaproic acid (EACA), 175
  effect on extinction coefficient, 169
  in γD isolation, 166, 167, 177
AMP (see Cyclic AMP)
Angiotensin, 4, 5
  I, 4
  II, 4
  as hapten, 4
  location of antigen site, 32
  structure–function relationship, 32
8-anilinonaphthalene-1-sulfonate (ANS), 147
  nonspecific binding, 147
Antibodies
  combining site size, 3, 26, 27
  dependence on antigen structure, 98
  induced by RNA, 106
  net charge, 98
  to adrenocorticotropic hormone, 3, 31, 36
  to α, β, and γ chains, 24, 30

Antibodies (Cont'd)
  to allotypes, 106
  to angiotensin, 31
  to bradykinin, 31
  to buried determinants, 98
  to cytochrome c, 12, 30, 39
  to DNP, 32
  to α-DNP-oligolysine, 34
  to ferredoxin, 8, 9
  to flagellin, 81, 82, 86
  to gastrin, 31, 36, 67
  to glucagon, 7, 8
  to γ-D-glutamic acid capsular polypeptide,
    2, 31
  to hemoglobin, 23, 24, 30
  to insulin, 10–11, 12, 31
  to lysozyme, 30, 31
  to myoglobin, 20, 23, 30
  to RNase (RNase-A), 15, 16, 17, 30
  to TMV, 30
  to TMVP, 26, 27, 37, 39
  synthesis, 106
Antibody
  combining sites, 145
    affinities, 28, 147
    amino acids, in, 148, 160–162
    contribution of light and heavy chains,
      145
    hydrophobic areas, 27, 33, 148
    involvement of heavy and light chains,
      148, 160
    of IgG, 147
    of IgM, 147
    reagents, 146
    size, 3, 26, 27, 147
    specificity, 146
    structure, 162
    variable regions, 151, 160–162
  formation
    delayed hypersensitivity, 91
    induction, 91

Antibody (*Cont'd*)
  forming cells
    Ig synthesis, 128
    recruitment, 128
Antigen
  ABO, 72
    α and β chains, 30
  adrenocorticotropic hormone, 3, 31, 36
  angiotensin II, 4
  complexed with RNA, 96, 98, 100–101,
    102, 109
  consumption by macrophages, 94, 95
  cytochrome *c*, 12
  ferredoxin, 8
  glucagon, 7, 8
  γ -D-glutamic acid capsular polypeptide, 2
  hemoglobin, 23
  histocompatability (*see* Histocompatabil-
    ity antigens)
  HL-A, 51, 55–60, 62–64, 66–74
  immunodominant group, 21, 22
  in lymph nodes, 104
  insulin, 10, 11, 13
  lysozyme, 18
  α-DNP-oligolysines, 32, 34
  ribonucleases, 15–17
  role in induction of immune response, 98
  size, 2, 3, 6, 9, 21
  structure–function relationships, 3, 5, 6,
    9, 8, 11–12, 13–22, 26, 28, 32
  superantigen, 96, 104, 106
  surface-bound to macrophages, 95
  tobacco mosaic virus (TMV), 25
  tobacco mosaic virus protein (TMVP), 28
Antigenic determinants
  characteristic of γD, 173
  competitive inhibition, 147
  conformation dependence, 23, 24, 29,
    30–31
  H, 79
  hydrophobic groups, 33, 34
  of ACTH, location, 32
  of α chains of hemoglobin, 2, 25
  of flagellin, 79, 80
  on angiotensin, location, 32
  on Fab, 173
  on gastrin, location, 32
  on hemoglobin, location, 24
  on TMV, location, 25
  on TMVP, location, 27
  secondary structure, 31
  sequence dependence, 31, 32
  size, 146, 147, 148
  thymus dependence, 81
Aromatic azide, chemistry, 157–159

p-(arsonic acid) benzenediazonium fluoro-
    borate, 154
Autoantibodies, 36

B cells
  function, 35, 36, 38, 39
  origin, 35
BADE, 156
BADL, 156
Biogel P-200, 169
  of γD serum, 169
Bradykinin
  as hapten, 5
  determinant size, 6
  structure–function relationships, 5, 6

Cell-mediated immunity (*see* Delayed hyper-
    sensitivity)
Cell receptors, 84, 89, 90, 91, 111, 141
Chaotropic agents, 55–56
Chromotography
  DEAE-cellulose, 166
  of γD, 166
Chymotrypsin, 161
Cycle of mitosis
  log growth, 123, 124, 126
  resting (G or stationary), 123, 124, 128,
    129, 136
  transition, 123, 126, 127, 128
Cyclic AMP, 129, 130, 131
Cyclohexamide, 119
Cytochrome *c*
  as antigen, 12
  heterotope sites, 34
  structure, 12
  structure–function relationship, 13–15
Cytosine arabinoside, 127

δ -chains
  amino acid content, 171
  carbohydrate content, 168, 171, 173
  in model, 175
  interchain disulfide bonds, 170, 171, 175
  intrachain disulfide bonds, 170, 171, 175
  partial specific volume, 168
  preparation, 169, 170
  molecular weight, 168
  N-terminal amino acids, 176
Delayed hypersensitivity
  carrier specificity, 37
  in tissue culture, 88

Delayed hypersensitivity (*Cont'd*)
 mediated by histocompatability antigens,
  63
 relationship to antibody production, 90
 role of T cells, 35, 38
 to ACTH, 36, 37
 to autologous sequences, 36
 to ferredoxin, 37
 to flagellin, 81, 82–83, 84, 90
 to glucagon, 37
 to haptens, 37
 to insulin, 36
 to peptide of TMVP, 29
 to tolerance, 91
Dendritic cells, antigen binding, 88
Diazoketone, chemistry, 156, 157
Dinitrophenol (DNP), affinity for antibody,
  147
DNA
 bouyant density, 135, 136, 140
 circular, 138, 140
 cytoplasmic, 131, 132
 linear, 138
 localization, 132
 membrane-associated (*see* Membrane-associated DNA)
 mycoplasmal, 140, 141
 nuclear, 136, 137, 138, 140, 141
 sedimentation, 136
 supercoiled, 132, 135, 137, 140, 141
 synthesis, 123, 124, 126, 127, 128, 136

Elastase, 161
Electrophoresis
 of γD serum, 167
 polyacrylamide, 168
 starch gel, 170

Fab
 affinity for hapten, 146
 as antigen binding site, 148
 association constants, 148
 carbohydrate in, 173, 175
 from γD, 173–175
 interchain disulfide bond, 175
 nonspecificity binding of ANS, 147
Fc
 carbohydrate in, 173, 175
 in cytoplasm, 118, 119, 120
 from γD, 173–175
 interchain disulfide bond, 175

Fc (*Cont'd*)
 membrane-associated, 114, 115, 117, 118,
  119, 120
 on lymphocytes, 115
Ferredoxin
 as antigen, 8, 9
 delayed sensitivity to, 37
 determinant size, 9
 structure–function relationships, 8, 9, 32
Flagellin, 78
 amino acid composition, 79
 antigenic determinants, 80
 as antigen, 78, 81, 82
 chemical modification, 80, 82, 83
 dissociation, 79
 H antigenic specificity, 79
 immunogenicity, 78, 81, 85–86
 in structure of flagellum, 78
 molecular weight, 78
 preparation, 78
 proteolytic fragmentation, 80, 82
 reaction with lymphocytes, 86
 regulator of lymphocytic function, 89
 repolymerization, 78
 specific determinants, 79
 tolerance to, 81, 82, 84, 85, 86, 90
Flagellin polymer
 as antigen, 81, 90
 definition, 78
 immunogenicity, 86
Flagellum
 carbohydrates, 78
 dissociation, 78
 electron microscopic view, 78
 flagellin subunits, 78
 H antigens, 79

γD
 antibody activity, 165, 177
 biological activities, 178
 biological significance, 178
 carbohydrate, 171, 175
 catabolism, 177, 178
 chromotography on
  Biogel P-200, 167
  immunoabsorbent, 167
  Sephadex G-200, 167
 DEAE-cellulose chromotography, 166
 effect of acid, 14, 169
 effect of heat, 169
 electrophoresis of, 167
 extinction coefficient, 169
 fragmentation with
  papain, 173, 174

D (*Cont'd*)
  pepsin, 174
  plasmin, 166, 174, 175
  trypsin, 173, 175
  in body fluids, concentration, 177
  in serum, concentration, 165, 166, 176
  instability, 165
  interchain disulfide bonds, 170, 175
  molecular weight, 168
  myeloma protein, 165
  protein nitrogen, 169
  rate of synthesis, 177
  reduction of, 169
  sedimentation constants, 167
  spontaneous fragmentation, 166, 173–175
  structural model, 175
Gastrin
  as hapten, 6, 36
  location of determinant, 31
  structure, 6
  structure–function relationship, 7
Glucagon
  delayed reactions to, 37
  structure–function relationship, 8
γ-D-glutamic acid (*see* Poly-γ-D-glutamic
    acid)

Heavy chains
  affinity labeled, 154, 156, 158
  allelic variants, 149, 152, 160
  peptides in combining site, 160, 161
  variable regions, 151, 152, 160
Hemoglobin
  α and β chains as antigens, 24
  as antigen, 23
  determinant, 23, 24
  structure, 22–23
  synthesis, 107
Heterotopes, 34
Histocompatability antigens
  and immunosurveillance, 52
  antibody to, 52
  biological activities, 63–67
  chemical nature, 71–73
  cultured cells, 57–58
  function, 52
  genetic loci, 51
  HL-A (*see* HL-A antigen)
  location, 51, 53
  molecular weight, 60, 67, 68, 72
  purification, 59–63
  solubilization with
    detergents, 55
    KCl, 55–56, 58–59, 60, 73
    low-intensity sound, 55, 62, 63, 73
    papain, 55

Histone, 128
HL-A antigen
  amino acid composition, 68, 69, 70, 72
  biological activities, 68
  carbohydrate, 67, 68, 72
  immunological potency, 66
  lipid, 67, 68
  molecular model, 73
  molecular weight, 68, 70, 72
  peptide maps, 70
  sedimentation rate, 70
  solubilization with
    KCl, 56, 58, 66, 69, 73
    papain, 55

IgA
  affinity-labeled, 155, 156, 157, 169
  myeloma (*see* Myeloma proteins)
IgG
  allotypes, 145
  as inhibitor, 114–115
  binding of ANS, 147
  Fc fragments, 114, 115
  on lymphocytes, 118
  on membranes, 115–117
  synthesis, 126–127
IgM
  membrane-associated, 114–115
Immunoabsorbent, 167
  isolation of γD, 167, 179
  preparation of, 167
Immunocompetent cell receptors, 29, 38
  specificity, 39
Immunoglobulin
  cytoplasmic, 119, 120, 122
  disappearance rates, 120, 122
  membrane-associated (M-Ig) (*see* Mem-
      brane-associated immunoglobulin and
      Specific Ig)
  synthesis, 107, 108, 115, 117, 120, 122,
      126, 128, 129
Immunosurveillance, concept, 52
Induction
  enzyme, 128
  of immune response, 128, 141
  of lymphocyte proliferation, 128, 141
Inhibition, competitive, 147–148
Insulin
  as antigen, 10
  delayed skin reactions to, 36
  structure, 10,
  structure–function relationships, 11–12

Landsteiner, antibodies to small molecules,
    145

Light chains
    affinity labeled, 154, 155, 156
    allelic variants, 149
    Bence–Jones proteins, 180
    γD types, 178
    in γD model, 175
    interchain disulfide bonds, 170
    κ, 115, 117, 118, 119, 120
    λ, 115
    membrane-associated, 114, 115, 118, 119
    M-RNA, 130
    peptides in combining sites, 159, 160
    separation from γD, 169, 170
    sequence variabilities, 150, 151
    synthesis, 129, 130
    variable regions, 151, 152, 161
Lymphocytes
    antigen binding, 118
    culture, 57–59
    cultured, 111–113, 120, 123
    density, 126
    differentiation, 112–113
    diploid, 111, 123, 138
    DNA, 123, 124, 132
    Fc fragments, 115
    fractionation, 94
    frequency of antigen binding, 89
    generation time, 112
    HL-A, in, 68
    IgG, 118
    induction, 128
    in peritoneal exudates, 93
    in primary response, 94
    interaction with dendritic cells, 85
    interaction with macrophages, 85
    κ chains, 115
    λ chains, 115
    membrane-associated Ig, 115, 118, 120
    morphology, 129
    M-RNA, 122–123, 129, 130
    mycoplasma, 140
    viral genome, 140–141
    polyribosomes, 122, 125–126
    reaction with antigen, 88–89, 90
    reaction with flagellin, 85–86, 88
    regulation by antigen, 89
    role in specific responses, 95
    transformation, 38, 63
Lysozyme, 162–164
    as antigen, 18
    cellular immune response to, 19
    heterotopes, 19
    structure–function relationships, 18–19

Macrophages, 38
    adherence to glass or plastic, 94

Macrophages (Cont'd)
    antigen binding, 88
    antigen catabolism, 94, 95
    effect of irradiation, 95
    function, 93, 94, 95, 96, 108
    induction of antibody production, 88
    induction of tolerance, 88
    in peritoneal exudates, 93
    in primary response, 94, 95
    ribonucleoprotein, 98
    RNA, 96, 108, 109
    role, 91
    surface-bound antigen, 95
Melanin, 128
Membrane
    -associated DNA (see Membrane-associated
        DNA)
    -associated RNA, 130–131
    Fc fragment, 115, 118, 119, 120
    histocompatability antigens, 52, 53, 56
    IgG, 115, 117
    hydrophobic bonds in, 56
    κ chains, 115, 118, 119, 120
    allotypic specificities on, 53
    structure, 53
    preparation, 134
    turnover, 119
Membrane-associated DNA, 131, 134–141
    bouyant density, 135, 136, 140
    electron microscopy, 138
    form, 138
    function, 141
    replication, 137–138
    sedimentation, 133, 136
    synthesis, 136
Membrane-associated immunoglobulin, 112,
        117
    as receptor, 111, 130, 133, 141
    fate, 120
    Fc, 118, 119, 120
    half-disappearance time, 118, 119, 120, 122
    on lymphocytes, 115, 133
    M-RNA, 122–123
    method, 113–115
    phenotype, 115
    synthesis, 117, 122
M-RNA
    for M-Ig, 122–123
    total cellular, 122–123
Mitosis (see Cycle of mitosis)
Myeloma proteins, 165, 167
    binding of ANS, 147
    γD, 165, 166, 168, 173, 178
    MOPC 315 IgA, 155, 156, 157, 159, 161
    sedimentation constants, 167
Myoglobin
    antiserum to, 23

Myoglobin (*Cont'd*)
  artificial, 20
  determinant size, 21
  immunodominant group 21, 22
  structure, 20
  structure–function relationships, 20–22, 31

NP-40, 132, 133, 135, 137, 140

peritoneal cell
  inhibition of migration, 29, 37
  ribonucleoprotein, 98, 100
  RNA rich extracts, 96, 101, 106
Phospholipase C, 133, 134
Polyacrylamide gel electrophoresis (PAGE),
    purification of HL-A antigens, 59–63, 69
Poly-γ-D-glutamic acid
  as hapten, 2
  capsular polypeptide, 2
Polyribosomes, of lymphocytes, 122, 123, 125–126
Primary responses
  to polymerized flagellin, 94
  to red blood cells, 94
Pronase, 134, 136
Protein synthesis, 126, 128, 129, 130
Puromycin, 118, 119, 120

Regulation, 123, 131
Ribonuclease B (RNase B), 15
Ribonuclease (RNase-A), 15
  structure–function relationships, 15, 16–17, 32
Ribonucleoprotein
  from macrophages, 98
  from peritoneal cells, 98
  molecular weight, 98
  RNA/protein, 98
RNA
  as gene modifier, 108, 109
  as messenger, 106, 107, 109
  as template for RNA-dependent DNA
    polymerase, 107–109
  complexed with antigen, 96, 100–103, 109
  -dependent DNA polymerase, 107, 108, 109
  in lymphoid cells, 108
  enhancing effect, 98–99
  from lymph nodes, 104–106

RNA (*Cont'd*)
  from plasma cells, 107
  specificity of, 100
  transfer of cellular immunity, 103, 104
  transfer of immunity, 93, 96, 104, 107

Secondary responses, in primed lympho-
    cytes, 94
Sephadex G-200, 167, 168
  of γD serum, 167
Sodium deoxycholate (DOC), 133, 134, 136

T cells
  function 35, 36, 38, 39
  interaction with B cells, 38, 39
  origin, 35
Theory, ontogeny of responsiveness, 52
Thymus, function, 35
Tobacco mosaic virus (TMV)
  as antigen, 25
  determinant location, 25
  structure, 25
  structure–function relationships, 26, 28, 33
Tobacco mosaic virus protein (TMVP)
  as antigen, 28
  delayed hypersensitivity to, 29
  structure–function relationships, 26, 28, 33
  subunits, 25
Tolerance
  high dose, 82, 86
  induction, 81, 82
  induction by solubile antigen, 94
  in tissue culture, 86
  low dose, 81, 82, 85, 86, 90
  role of macrophages, 88
  to flagellin, 81, 84, 85, 86, 88, 90, 91
  to polymer, 81, 90
Transplantation antigens (*see* Histocompata-
    bility antigens)
T-RNA, forms in cytoplasm, 102
Trypsin, 161
Tyrosine aminotransferase, 128

van Krogh equation, 64
Variable regions
  allotype-related sequences, 150
  hypervariable region, 151
  in antibody combining site, 151
  in Fab, 149